AF723869

Biodiversity of Marine Microbes

Biodiversity of Marine Microbes

Editor

Savvas Genitsaris

MDPI • Basel • Beijing • Wuhan • Barcelona • Belgrade • Manchester • Tokyo • Cluj • Tianjin

Editor
Savvas Genitsaris
National and Kapodistrian
University of Athens
Greece

Editorial Office
MDPI
St. Alban-Anlage 66
4052 Basel, Switzerland

This is a reprint of articles from the Special Issue published online in the open access journal *Diversity* (ISSN 1424-2818) (available at: http://www.mdpi.com).

For citation purposes, cite each article independently as indicated on the article page online and as indicated below:

LastName, A.A.; LastName, B.B.; LastName, C.C. Article Title. *Journal Name* **Year**, *Volume Number*, Page Range.

ISBN 978-3-0365-1052-1 (Hbk)
ISBN 978-3-0365-1053-8 (PDF)

Cover image courtesy of Maria Moustaka-Gouni.

© 2021 by the authors. Articles in this book are Open Access and distributed under the Creative Commons Attribution (CC BY) license, which allows users to download, copy and build upon published articles, as long as the author and publisher are properly credited, which ensures maximum dissemination and a wider impact of our publications.
The book as a whole is distributed by MDPI under the terms and conditions of the Creative Commons license CC BY-NC-ND.

Contents

About the Editor

Savvas Genitsaris has a Ph.D. in Biodiversity and Molecular Ecology from Aristotle University of Thessaloniki, Greece. He worked as a post-doctoral researcher in the University of Littoral (ULCO) in France, where he focused his research on the investigation of biotic and abiotic interactions of marine microbes with the use of metagenomics and bioinformatics tools. He was a post-doctoral researcher of the International Hellenic University and in Aristotle University of Thessaloniki, where he worked on coastal algal blooms and red tides. Currently he is an Assistant Professor in the National and Kapodistrian University of Athens, Greece, where his research focuses on the diversity and function of aquatic microbial communities of the three Domains of Life. His interests include the exploration of the biodiversity of microorganisms in freshwater, marine, coastal and brackish systems, using microscopy, metagenomics, DNA and RNA sequencing, bioinformatics and ecological tools. He is also interested in examining the biotic and abiotic relationships that shape aquatic microbial communities.

Preface to "Biodiversity of Marine Microbes"

Marine microbial life is comprised of a variety of different evolutionary groups from all three domains of life, Eukaryotes, Bacteria, and Archaea. It is responsible for about half of the primary production on earth, plays irreplaceable roles in biogeochemical cycles and ecosystem functioning, and actively participates in complex processes and interactions. Marine microbes are the basis of marine trophic webs (autotrophs), and an important link between different trophic levels, as decomposers, parasites, and endosymbionts. They are used as biological indicators of water quality, eutrophication, and degraded marine environments, and are targeted in conservation and restoration plans. Our understanding of their responses to climate change is considered a key research field to comprehend the complex ongoing processes that will shape the planet's future. They consist of a vast diversity of organisms, with diverse morphological features, sizes, physiology, functions, trophic characteristics, distribution, ecology, evolutionary traits, genetic content, and responses to abiotic variability. Even though we understand their significance in numerous aspects that affect all life on earth, we still have a long way to go to answer fundamental questions driving recent research: How many marine microbes are there? Where can we find them? What do they do? What is their role and responses in light of climate change? What are their phylogenetic relationships? How do they respond to environmental pressures? This along with many more. Advances in high-throughput sequencing accompanied with the technological innovations of classical tools, such as microscopy, have given researchers the equipment and the incentive to attempt to tackle the above questions and shed light on the complex and diverse life of marine microbes. This book provides a platform to highlight new research and significant advances related to the biodiversity of marine microbes.

Savvas Genitsaris
Editor

diversity

MDPI

Editorial

Biodiversity of Marine Microbes

Savvas Genitsaris

School of Science and Technology, International Hellenic University, 57001 Thermi, Greece; s.genitsaris@ihu.edu.gr

Received: 15 June 2020; Accepted: 15 June 2020; Published: 16 June 2020

Abstract: The Special Issue entitled "Biodiversity of Marine Microbes" aimed at highlighting the significance of marine microbes as primary producers, their participation in complex processes and interactions with both the biotic and the abiotic environment, and their important roles in biogeochemical cycles and ecosystem functioning. The issue includes five research papers, covering the diversity and composition of marine microbial communities representing all three domains of life in various marine environments, including coastal eutrophic areas, ice waters, and lagoons. One paper examined the diversity and succession of bacterial and archaeal communities from coastal waters in mesocosm experiments. The combination of classical tools with novel technological advances offers the opportunity to answer fundamental questions and shed light on the complex and diverse life of marine microbes.

Keywords: algae; prokaryotes; microeukaryotes; coastal eutrophic systems; high-throughput sequencing; blooms; climate change

---

Marine microbial communities comprise a vast diversity of different evolutionary groups from all three domains of life. Recent technological advances in high-throughput sequencing combined with innovations in classical tools, such as microscopy, have permitted the in-depth examination of these communities, and have advanced our knowledge on the key roles that these microbes may play in marine systems. It is well established that marine algae contribute half of the planet's primary production, microeukaryotes are key players as top-down heterotrophs on marine trophic webs, and bacteria and archaea are essential links in global biogeochemical cycles and marine ecosystem functioning. However, many questions remain underexplored, which are relevant, for example, to the response of coastal protistan communities in high eutrophication conditions, the diversity of picoeukaryotes in extreme environments such as ice waters, the phytoplankton composition along variable conductivity gradients, the effects of hydrography on the marine microbial communities' structure, and the changes in picoplankton under different light conditions and bloom events.

This Special Issue comprises five research articles attempting to tackle the above-mentioned subjects of the biodiversity of marine microbes. Genitsaris et al. investigated the plankton community's composition and abundance in an urban eutrophic coastal area of the Mediterranean, amid frequent and persistent phytoplankton blooms, red tide events, mucilaginous aggregates, and the proliferation of potentially harmful species. This paper provided evidence of the eutrophication effects on the response of eukaryotic plankton assemblages and their impact on water quality and ecosystem services [1]. A metagenomic study of the under-ice photosynthetic picoeukaryotes of the White Sea basin, by means of 18S rRNA gene high-throughput sequencing, indicated that environmental variability in extreme marine environments could lead to a significant shift in microbial communities' composition and structure [2]. In addition, a study of phytoplankton diversity patterns along salinity variations in Mediterranean lagoons suggested that the communities' heterogeneity was highly associated with the recorded differences in conductivity among the different sites [3]. Similarly, Gong et al. showed that spatial hydrographic variability in Taiwan Strait lead to significant variations in all the components of the microbial food web, including viruses, picoplankton, nanoflagellates and ciliates, and showed that, during the cold months, a "viral loop" might contribute to the top-down control of bacterial populations [4]. Finally, one paper in the Special Issue concerned a mesocosm experimental set-up, aiming to investigate bacterial and archaeal diversity and succession in changing light regimes during phytoplankton growth. This study confirmed that light irradiance can affect marine bacterial communities' structure both directly and indirectly, which in turn can have significant implications for a marine ecosystem's response to environmental change [5].

In my view, and according to the published information in this Special Issue, the main challenge of the coming years and the main goal of the future marine microbiologists, is to successfully combine novel technological advances and integrate the recent sequencing breakthroughs with the classical analytical tools, in order to reveal the true vast diversity of marine microbes and to understand the structure of these communities in relation to abiotic disturbances, environmental pressures and biotic relations.

Conflicts of Interest: The author declares no conflict of interest.

References

1. Genitsaris, S.; Stefanidou, N.; Sommer, U.; Moustaka-Gouni, M. Phytoplankton blooms, red tides and mucilaginous aggregates in the urban Thessaloniki Bay, Eastern Mediterranean. *Diversity* **2019**, *11*, 136. [CrossRef]
2. Belevich, T.A.; Ilyash, L.V.; Milyutina, I.A.; Logacheva, M.D.; Trotsky, A.V. Photosynthetic picoeukaryotes diversity in the underlying ice waters of the White Sea, Russia. *Diversity* **2020**, *12*, 93. [CrossRef]
3. Stefanidou, N.; Katsiapi, M.; Tsianis, D.; Demertzioglou, M.; Michaloudi, E.; Moustaka-Gouni, M. Patterns in Alpha and Beta Phytoplankton Diversity along a Conductivity Gradient in Coastal Mediterranean Lagoons. *Diversity* **2020**, *12*, 38. [CrossRef]
4. Gong, G.-C.; Yeh, H.-M.; Chen, Y.-K.; Hsieh, C.-H.; Ho, P.-C.; Tsai, A.-Y. Effect of hydrographic variability on the distribution of microbial communities in Taiwan Strait in winter. *Diversity* **2019**, *11*, 193. [CrossRef]
5. Karayanni, H.; Kormas, K.A.; Moustaka-Gouni, M.; Sommer, U. Changes in heterotrophic picoplankton community structure after induction of a phytoplankton bloom under different light regimes. *Diversity* **2019**, *11*, 195. [CrossRef]

© 2020 by the author. Licensee MDPI, Basel, Switzerland. This article is an open access article distributed under the terms and conditions of the Creative Commons Attribution (CC BY) license (http://creativecommons.org/licenses/by/4.0/).

diversity

MDPI

Article

Phytoplankton Blooms, Red Tides and Mucilaginous Aggregates in the Urban Thessaloniki Bay, Eastern Mediterranean

Savvas Genitsaris [1,2], Natassa Stefanidou [1], Ulrich Sommer [3] and Maria Moustaka-Gouni [1,*]

1 Department of Botany, School of Biology, Aristotle University of Thessaloniki, 54124 Thessaloniki, Greece
2 School of Economics, Business Administration and Legal Studies, International Hellenic University, 57001 Thermi, Greece
3 Geomar Helmholtz Centre for Ocean Research Kiel, 24105 Kiel, Germany
* Correspondence: mmustaka@bio.auth.gr; Tel.: +30-2310-998325

Received: 5 July 2019; Accepted: 12 August 2019; Published: 14 August 2019

Abstract: We investigated the plankton community composition and abundance in the urban marine environment of Thessaloniki Bay. We collected water samples weekly from March 2017 to February 2018 at the coastal front of Thessaloniki city center and monthly samples from three other inshore sites along the urban front of the bay. During the study period, conspicuous and successive phytoplankton blooms, dominated by known mucilage-producing diatoms alternated with red tide events formed by the dinoflagellates *Noctiluca scintillans* and *Spatulodinium pseudonoctiluca*, and an extensive mucilage aggregate phenomenon, which appeared in late June 2017. At least 11 known harmful algae were identified throughout the study, with the increase in the abundance of the known harmful dinoflagellate *Dinophysis* cf. *acuminata* occurring in October and November 2017. Finally, a red tide caused by the photosynthetic ciliate *Mesodinium rubrum* on December 2017 was conspicuous throughout the sampling sites. The above-mentioned harmful blooms and red tides were linked to high nutrient concentrations and eutrophication. This paper provides an overview of eutrophication impacts on the response of the unicellular eukaryotic plankton organisms and their impact on water quality and ecosystem services.

Keywords: nutrients; HABs; mucilaginous aggregates; *Noctiluca scintillans*; *Dinophysis*; *Mesodinium rubrum*

1. Introduction

On a global scale, the rate of coastal urbanization will increase rapidly in the next decades, and in combination with climate change is projected to result in an increased risk of coastal eutrophication [1,2]. Sewage inputs from coastal cities that are transported directly to coastal waters can act synergistically with land-based sources and river run-off causing high levels of nutrients [3,4]. Consequently, the global Indicator for Coastal Eutrophication Potential (ICEP) analyses indicate that the potential for coastal eutrophication continuously grows worldwide [2]. Worldwide eutrophication has led to phytoplankton abundance and biomass increase [5–7], while more coastal harmful algal blooms (HABs), with more toxic species, have been linked with eutrophication phenomena [8,9]. Numerous examples of linkages between nutrient loading and coastal phytoplankton blooms and mucilagine aggregate phenomena [10,11] include the involvement of harmful species, i.e., the diatom *Pseudonitzschia* spp. in the Gulf of Mexico [12], the dinoflagellates *Prorocentrum* sp., and *Karenia mikimotoi* along the coast of China [13], and the red tide-forming heterotrophic dinoflagellate *Noctiluca scintillans* [14–16].

A large volume of domestic and industrial wastes from the city of Thessaloniki has been directed for decades to the Thermaikos Gulf and especially its inner part, Thessaloniki Bay. In the 20th century, these wastes were discharged in the Bay without treatment, causing the eutrophication of the system. Since 2001, wastewater treatment has been implemented, decreasing the effects of anthropogenic eutrophication [17]. Although it is generally accepted in the public that the water quality in the Thermaikos Gulf has been improved compared to 20–30 years ago [18,19], the urban front of the Thessaloniki Bay, with restricted water circulation and shallowness, still exhibits apparent red tides and algal blooms. These events are usually the cause of irritation to the public, often mentioned in the Greek media, with subsequent socio-economic consequences to the city of Thessaloniki, especially the touristic center. Despite the growing concerns of the citizens and authorities on the water quality of the Bay and particularly of the urban front, only scarce and isolated studies have been published on the abundance and dynamics of plankton community (both phyto- and protozooplankton) in the urban part of the Gulf [20,21]. On the other hand, several studies in the broader Thermaikos Gulf have focused on phytoplankton [18,22,23] and the occurrence of HABs [24,25]. However, comprehensive studies on red tides and mucilage aggregate phenomena are lacking for the Thermaikos Gulf.

According to the related legislation for the ecological water quality based on nutrient pressures and the phytoplankton quality element (Water Framework Directive; WFD, 2000/60/EC) there is a need for systematic and frequent monitoring of coastal waters. Furthermore, similar to the WFD objectives are those of Marine Strategy Framework Directive (MSFD, 2008/56/EC) for achieving good environmental status of EU marine waters by 2020. The MSFD eutrophication quality descriptor (D5) refers to the adverse effects of eutrophication including harmful algae blooms [26].

The aim of this paper was to examine the shift in the protagonists of the conspicuous and successive algal blooms, red tides and mucilage aggregations in the urban marine environment of the Thessaloniki Bay, by investigating the temporal and spatial changes of the unicellular eukaryotic plankton community attributes (species diversity, dominance, and abundance). This is the first study concerning the phyto- and protozoo-plankton species succession at an annual time scale with weekly samplings, with conspicuous phytoplankton blooms, red tides and a mucilage aggregate phenomenon, in the urban coastal front of the Thessaloniki Bay (Thermaikos Gulf). The present work focuses on the zone linking the terrestrial and the coastal environments heavily affected by, and influencing various human activities, such as the harbor, tourism, industry, mussel cultures, and sewage effluents.

2. Materials and Methods

2.1. Sampling Sites and Sample Collection

Samples were collected weekly from March 2017 to February 2018, from a coastal inshore sampling site in the White Tower (WT) in the center of the city of Thessaloniki (Table 1; Figure 1). During the same period, every month, additional samples were collected from three other inshore sites along the urban front of the Bay, namely at Aretsou Beach (AR), Music Hall coast (MH), and harbor (HB) (Table 1; Figure 1). In total, 47 samples were collected from WT and 12 from each other site (Table 1). All sampling sites had a maximal depth of 4 m.

Table 1. Sampling sites and total number of samples collected.

Sampling Sites	Latitude	Longitude	Number of Samples
White Tower (WT)	40°37′34 N	22°56′51 E	47
Aretsou Beach (AR)	40°34′29 N	22°56′38 E	12
Music Hall coast (MH)	40°35′57 N	22°56′53 E	12
Harbor (HB)	40°37′55 N	22°56′09 E	12

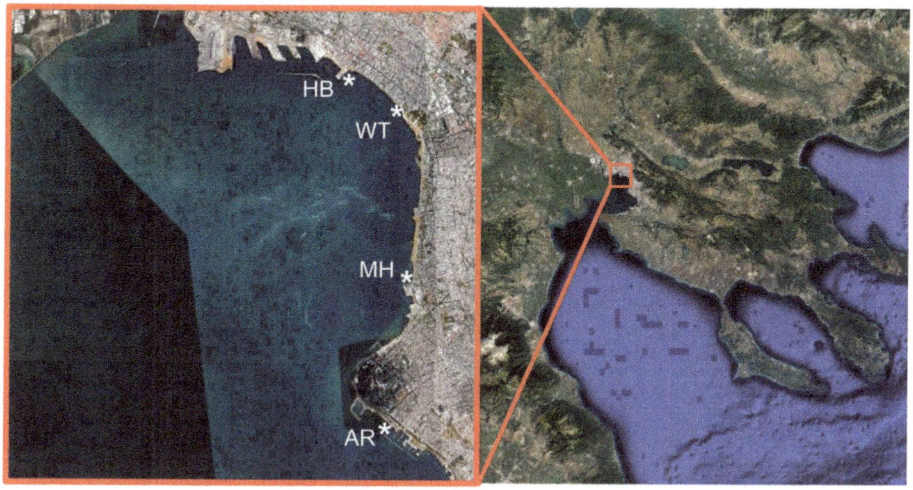

Figure 1. Study area in Thermaikos Bay, indicating the location of the four sampling sites (*). WT: White Tower, AR: Aretsou, MH: Music Hall, HB: Harbor.

During all samplings, in situ measurements of water temperature and conductivity were made with the use of the YSI Pro 1030 instrument (YSI Inc., Yellow Springs, OH, USA). Conductivity was transformed to salinity based on the equation in Weyl [27]. Water samples of 2 L were collected from the surface layer of 1 m, and separated as follows: (i) a subsample of 0.5 L was used for immediate microscopic observation of the living microbial eukaryotic community; (ii) a subsample of 0.5 L was preserved in Lugol's solution and kept in the dark in room temperature for microscopic analysis within the next few days; (iii) subsamples of 100–250 mL (depending on plankton and particulate matter density) were immediately filtered onto 0.7 μm pre-washed (in 5–10% HCl) and pre-combusted (6 h, 550 °C) Whatman GF/F filters, and the filters were stored in −20 °C until future particulate organic phosphorus and chlorophyll *a* (Chl *a*) measurements; (iv) a subsample of 50 mL was filtered through 0.2 μm cellulose acetate filters (Sartorius) and the filtered water aliquots were kept in −20 °C until future dissolved inorganic nutrient measurements.

2.2. *Chl a and Nutrient Measurements*

Chl *a* content was estimated according to Jeffrey and Humphrey [28]. Prior to the photochemical measurements (HITACHI, U2900) filters were put into 8 mL acetone (90%) for 24 h in the dark at 6 °C.

Particulate organic phosphorus (POP) was measured colorimetrically by an element analyzer (Thermo Scientific Flash 2000) at 882 nm, following the protocol by Hansen and Koroleff [29]. Nitrate and nitrite (NO_3^- and NO_2^-), ammonium (NH_4^+), silicate (SiO_4^-), and phosphate (PO_4^-) were, also, measured according to Hansen and Koroleff [29].

Furthermore, the Eutrophication Index (E.I.) of Primpas et al. [30] was used in order to assess the eutrophication status of Thessaloniki Bay. The formula takes into consideration the NO_3^- and NO_2^-, ammonium, PO_4^-, and Chl *a* concentrations resulting in three distinct ranges describing oligotrophy (0.04–0.38), mesotrophy (0.37–0.87), and eutrophication (0.83–1.51). The ranges are further divided into a five-scale scheme according to the WFD requirements, in order to assess the water quality status, as follows:

1. High ecological water quality: <0.04
2. Good: 0.04–0.38
3. Moderate: 0.38–0.85
4. Poor: 0.85–1.51

5. Bad: >1.51

The E.I. was calculated according to the following equation:

$$E.I. = 0.279 \times PO_4 + 0.261 \times NO_3 + 0.296 \times NO_2 + 0.275 \times NH_3 + 0.214 \times Chl\ a$$

2.3. Microscopic Analysis

Planktonic unicellular eukaryotes were examined in sedimentation chambers using an inverted epi-fluorescence microscope (Nikon Eclipse TE 2000-S, Melville, USA), with phase contrast. Taxa were identified based on taxonomic keys and relevant papers [31–33]. Light and phase-contrast micrographs of live and Lugol-preserved cells were taken using a digital microscope camera (Nikon DS-L1, Melville, USA). Plankton counts (cells and colonies) were performed using the inverted microscope method [34]. At least 400 individuals in total, and 100 individuals of the most abundant taxa, were counted per sample in sedimentation chambers. Taxa comprising of > 10% of the total plankton abundance per sample were arbitrarily considered to be dominant for that particular sample. Population density of 1000 cells mL^{-1} for a particular phytoplankton taxon in a sample was considered as a baseline bloom density in this urban coastal environment. This threshold is based on the Greek eutrophication scale and the total phytoplankton abundance (960 cells mL^{-1}) given as an indicator of bad water quality or eutrophic coastal waters [35]. Potentially harmful plankton taxa identified during the study, were acknowledged according to the IOC-UNESCO Taxonomic Reference List of Harmful Microalgae.

2.4. Data Analysis

Alpha-diversity estimators (the Simpson, Shannon, Evenness, Equitability, and Berger–Parker indices) were calculated with the PAST 2.17c software [36] in all samples. These indices have been reported to better describe general properties of the communities [37] and reflect anthropogenic or environmental variability effects on ecosystem functions [38]. Paired t-tests were applied in PAST 2.17c software to compare the (i) physical and chemical variables and (ii), richness, abundance, and the alpha-diversity estimators between the four sampling sites. The p-values < 0.05 indicated significant differences between pairwise comparisons. Furthermore, pairwise comparisons of sampling sites, based on the relative abundance of individual taxa, were implemented with the Kolmogorov–Smirnov test in the PAST 2.17c software.

The plankton assemblages of the different samplings were compared using the Plymouth routines in the multivariate ecological research software package PRIMER v.6 [39]. The Jaccard coefficients were calculated to develop the matrix based on taxa abundance in order to identify interrelationships between samples and construct cluster and MDS (multi-dimensional scaling) plots. The similarity profile (SIMPROF) permutation test was conducted to determine the significance of the dendrogram branches resulting from cluster analysis.

Network analysis was performed in order to explore strong relationships among plankton taxa, and between plankton taxa and environmental parameters in all samplings. The relationships were characterized through MINE (maximal information-based nonparametric exploration) statistics by computing the maximal information coefficient (MIC) between each pair of taxa, and pairs of taxa and environmental parameters [40], considering abundance values for each taxon. MIC is a non-parametric method which captures associations (linear or non-linear) between data pairs. It provides a score that represents the strength of the relationship. The matrix of MIC values corresponding to p-values < 0.01, based on pre-computed p-values of various MIC scores at different sample sizes, was used to visualize the networks of associations with Cytoscape 3.5.1 [41]. Furthermore, correlation analysis was conducted in order to investigate the relationships between the plankton taxonomic groups and the E.I., and the inorganic nutrient molar ratios (N:P, Si:N, Si:P).

3. Results

3.1. Environmental Parameters

Seawater temperature recorded in the WT sampling site during the period of the study ranged from 9.60 to 29.7 °C and salinity from 32.8 to 38.8 (Table 2), mean 37.2. The concentrations of inorganic nutrients (SiO_4, PO_4, NO_3, NO_2, NH_4), and particulate organic phosphorus (POP) all exhibited strong fluctuations throughout the entire study period, with some extreme high values recorded in all sites, especially in WT (Supplementary Figure S1). In particular, remarkably high values were recorded for all nutrients on 22 March 2017 in the WT, (SiO_4: 10.17 μmol L^{-1}; PO_4: 9.54 μmol L^{-1}; NO_2: 0.77 μmol L^{-1}; NO_3 and NO_2: 7.53 μmol L^{-1}; NH_4: 160.3 μmol L^{-1}; POP: 42.1 μmol L^{-1}), followed by high values of most nutrients during the sampling on 28 June 2017 in WT and MH, 20 September 2017 in all sites, and 10 January 2018 in WT. Additionally, the highest value of NH_4 (32.86 μmol L^{-1}) was recorded on 18 October 2017 in HB. The annual means of Chl *a* for each station were as follows: WT: 2.62 μg L^{-1}, AR: 3.11 μg L^{-1}, MH: 1.43 μg L^{-1} and HB: 3.15 μg L^{-1}. Furthermore, Chl *a* showed marked variability, ranging from 0.27 μg L^{-1} on 27 December 2017 in WT, to 17.28 μg L^{-1} on 13 December 2017 in WT. The mean inorganic nutrient ratios were 25.1 for N:P, 0.70 for Si:N and 18.4 for Si:P. The maximum N:P (71.9) and Si:P (76.6) ratios in WT was measured on 28 June 2017 while the minimum N:P ratio (10.1) on 13 December 2017 and the minimum Si:P ratio (1) on 22 March 2017. In the rest of the sites, the maximum N:P and Si:P ratios were recorded in August samples and the minimum in November (HB) and December samples (AR, MH). The calculated Si:N ratios were relatively low (< 2) in all samples with maximum values coinciding with diatom blooms. All the calculated E.I. values exceeded the value 0.83 in all samples (data not shown), thus were indicative for eutrophication, reaching the highest value during the 22 March 2017 red tide event (49.5).

Considering the common sampling dates conducted in the different sampling sites, no significant differences were found in almost all paired comparisons of environmental parameters, based on t-tests (for a visualization of mean values of environmental parameters in each sampling site, see Supplementary Figure S2). Significant differences were found between sites for NO_2 (with AR > MH), for NO_3 and NO_2 (WT > MH; AR > MH), and for POP (WT > MH; MH < HB) (Supplementary Table S1). Furthermore, higher ammonia (NH_4) concentrations were recorded at HB (Table 2), even though no statistically significant differences between sites were found.

Table 2. Sample dates, sites and coding, and values of abiotic parameters (water temperature, salinity, SiO_4, PO_4, NO_2, NO_3 and NO_2, NH_4, POP – Particulate Organic Phosphorus, Chl *a*). All nutrient concentration values are given in μmol L^{-1}. The sampling sites are shown in Figure 1.

Sample Date and Site	Sample Code	Temperature of Water (°C)	Salinity	SiO_4	PO_4	NO_2	NO_3 & NO_2	NH_4	POP	Chl *a* (μg L^{-1})
1. 15 March 2017–W. Tower	15MarWT	12.1	38.8	3.85	0.75	0.37	3.59	7.59	0.20	0.60
2. 22 March 2017–W. Tower	22MarWT	14	38.6	10.17	9.54	0.77	7.53	160.3	42.1	3.54
3. 29 March 2017–W. Tower	29MarWT	14.1	37.2	4.98	0.42	0.09	1.72	3.55	0.46	1.67
4. 05 April 2017–W. Tower	05AprWT	16.4	37.3	6.33	0.41	0.40	2.28	6.71	0.85	2.42
5. 12 April 2017–W. Tower	12AprWT	15.5	37.7	8.49	0.78	0.08	8.36	11.2	2.81	2.76
6. 12 April 2017–Aretsou	12AprAR	16.1	37.3	3.79	0.27	0.20	2.52	4.48	0.48	1.39
7. 12 April 2017–Music Hall	12AprMH	15.4	38.3	3.94	0.30	0.06	1.00	3.04	0.30	1.29
8. 12 April 2017–Harbour	12AprHB	15.5	38.2	4.52	0.25	0.16	1.39	3.55	0.78	2.79

Table 2. *Cont.*

Sample Date and Site	Sample Code	Temperature of Water (°C)	Salinity	SiO_4	PO_4	NO_2	NO_3 & NO_2	NH_4	POP	Chl *a* (μg L^{-1})
9. 19 April 2017–W. Tower	19AprWT	15.8	38	2.59	0.19	0.04	0.54	1.99	0.21	0.60
10. 26 April 2017–W. Tower	26AprWT	15.1	38.2	3.36	0.35	0.10	1.82	4.36	0.28	0.78
11. 03 May 2017–W. Tower	03MayWT	19	37.3	3.63	0.22	0.19	2.69	3.53	0.38	1.14
12. 09 May 2017–W. Tower	09MayWT	18.6	37.8	4.76	0.20	0.15	7.36	4.61	0.25	0.95
13. 09 May 2017–Aretsou	09MayAR	19.5	38	4.10	0.18	0.21	2.26	4.63	0.22	1.10
14. 09 May 2017–Music Hall	09MayMH	19	38.1	2.37	0.30	0.13	1.71	4.45	0.14	0.48
15. 09 May 2017–Harbour	09MayHB	18.3	38.1	2.50	0.16	0.15	1.54	3.73	0.23	0.71
16. 17 May 2017–W. Tower	17MayWT	20.1	37.1	6.34	0.49	0.19	7.08	5.32	0.97	2.77
17. 24 May 2017–W. Tower	24MayWT	22	32.8	2.95	0.15	0.18	1.97	2.67	0.49	3.63
18. 31 May 2017–W. Tower	31MayWT	22.7	37.4	2.55	0.18	0.21	5.03	2.03	0.62	4.70
19. 07 June 2017–W. Tower	07JunWT	25.4	36.7	3.44	0.57	0.27	2.36	8.08	0.55	1.40
20. 07 June 2017–Aretsou	07JunAR	25.7	36.3	5.93	0.20	0.14	4.70	2.96	0.51	2.17
21. 07 June 2017–Music Hall	07JunMH	24.9	36.7	4.13	0.43	0.10	0.89	4.36	0.48	1.04
22. 07 June 2017–Harbour	07JunHB	25.9	36.5	4.87	0.48	0.21	3.11	6.01	0.49	1.14
23. 14 June 2017–W. Tower	14JunWT	25.5	36.7	6.24	0.40	0.33	3.91	7.53	0.54	1.74
24. 21 June 2017–W. Tower	21JunWT	23.1	36.2	4.13	0.27	0.26	2.55	5.10	0.26	1.01
25. 28 June 2017–W. Tower	28JunWT	28	36.1	18.39	0.24	0.18	10.93	6.33	1.87	6.85
26. 28 June 2017–Aretsou	28JunAR	28.4	36.2	6.99	0.18	0.14	3.72	3.34	0.70	4.10
27. 28 June 2017–Music Hall	28JunMH	27.9	36	10.04	0.44	0.21	4.46	6.99	0.38	1.11
28. 28 June 2017–Harbour	28JunHB	28.7	36.4	9.44	0.14	0.07	2.48	2.37	0.86	4.44
29. 04 July 2017–W. Tower	04JulWT	20.5	35.7	6.79	0.49	0.16	2.43	7.68	0.23	1.21
30. 12 July 2017–W. Tower	12JulWT	28	36.7	4.69	0.19	0.07	1.98	4.24	0.41	2.56
31. 19 July 2017–W. Tower	19JulWT	23.4	35.3	5.24	0.15	0.09	1.32	2.13	0.91	9.90
32. 26 July 2017–W. Tower	26JulWT	28.8	35.9	6.42	0.14	0.07	1.58	2.36	0.56	1.98
33. 26 July 2017–Aretsou	26JulAR	28.1	36.4	5.47	0.12	0.20	3.34	2.47	0.58	1.75
34. 26 July 2017–Music Hall	26JulMH	29.3	36	5.25	0.13	0.10	0.72	1.93	0.79	2.14
35. 26 July 2017–Harbour	26JulHB	29	35.5	6.65	0.14	0.07	0.88	4.45	0.67	1.24
36. 02 August 2017–W. Tower	02AugWT	27.2	36.3	7.33	0.25	0.11	2.34	4.96	0.31	0.79
37. 09 August 2017–W. Tower	09AugWT	29.7	36.6	7.91	0.30	0.15	7.21	7.26	0.66	1.13

Table 2. *Cont.*

Sample Date and Site	Sample Code	Temperature of Water (°C)	Salinity	SiO_4	PO_4	NO_2	NO_3 & NO_2	NH_4	POP	Chl *a* (μg L^{-1})
38. 23 August 2017–W. Tower	23AugWT	24.6	37.7	6.00	0.24	0.09	1.01	3.50	0.25	0.77
39. 23 August 2017–Aretsou	23AugAR	25	36.8	7.08	0.24	0.25	2.21	3.03	0.29	1.05
40. 23 August 2017–Music Hall	23AugMH	24.9	37.5	5.91	0.21	0.08	1.99	3.11	0.21	1.04
41. 23 August 2017–Harbour	23AugHB	24.6	37.7	N/A	N/A	N/A	N/A	N/A	0.20	0.77
42. 30 August 2017–W. Tower	30AugWT	24.9	36.3	6.03	0.63	0.40	6.29	9.63	0.74	1.86
43. 06 September 2017–W. Tower	06SepWT	24.8	37.6	8.70	0.30	0.19	3.53	6.00	0.78	1.23
44. 13 September 2017–W. Tower	13SepWT	26.2	37.2	3.10	0.17	0.16	1.27	3.51	0.43	1.81
45. 20 September 2017–W. Tower	20SepWT	26.6	37.3	11.83	0.25	0.19	3.80	6.64	0.84	1.66
46. 20 September 2017–Aretsou	20SepAR	26.4	37.3	10.08	0.16	0.46	5.75	5.68	0.40	2.25
47. 20 September 2017–Music Hall	20SepMH	26.8	37.4	13.37	0.34	0.20	2.54	8.54	0.66	1.50
48. 20 September 2017–Harbour	20SepHB	26.6	37.3	13.14	0.27	0.23	4.78	7.04	0.42	1.60
49. 27 September 2017–W. Tower	27SepWT	22.6	37.2	1.27	0.28	0.08	1.44	2.16	0.16	2.74
50. 04 October 2017–W. Tower	04OctWT	22.2	37.4	2.78	0.38	0.14	2.16	3.18	0.16	0.66
51. 11 October 2017–W. Tower	11OctWT	20.9	37.4	2.31	0.32	0.29	2.11	4.53	0.17	0.72
52. 18 October 2017–W. Tower	18OctWT	20.8	38.1	2.60	0.31	0.08	1.92	3.14	0.28	1.29
53. 18 October 2017–Aretsou	18OctAR	20.5	37.4	2.07	0.22	0.09	1.80	2.49	0.21	2.24
54. 18 October 2017–Music Hall	18OctMH	20.8	37.6	1.25	0.23	0.07	0.79	2.69	0.20	1.08
55. 18 October 2017–Harbour	18OctHB	20.8	37.7	2.66	3.71	0.19	2.03	32.86	1.12	1.50
56. 25 October 2017–W. Tower	18OctWT	19.8	36.7	2.39	0.32	0.11	4.41	3.40	0.17	0.98
57. 01 November 2017–W. Tower	01NovWT	17.4	37.6	2.94	0.35	0.19	2.10	3.29	0.17	0.71
58. 08 November 2017–W. Tower	08NovWT	16.7	37.6	1.35	0.30	0.10	1.02	2.92	0.39	0.94
59. 15 November 2017–W. Tower	15NovWT	16.7	37.5	2.98	0.30	0.18	3.82	6.38	0.77	6.58
60. 15 November 2017–Aretsou	15NovAR	16.9	37.7	2.50	0.31	0.21	2.92	3.54	0.19	1.10
61. 15 November 2017–Music Hall	15NovMH	16.8	37.8	2.77	0.23	0.11	1.85	3.09	0.08	1.34

Table 2. *Cont.*

Sample Date and Site	Sample Code	Temperature of Water (°C)	Salinity	SiO_4	PO_4	NO_2	NO_3 & NO_2	NH_4	POP	Chl *a* (μg L^{-1})
62. 15 November 2017–Harbour	15NovHB	16.6	37.3	1.07	0.30	0.08	1.23	7.54	1.50	10.05
63. 22 November 2017–W. Tower	22NovWT	15.2	37.8	1.32	0.27	0.10	0.72	2.42	0.15	1.02
64. 29 November 2017–W. Tower	29NovWT	13.7	37.7	5.05	0.51	0.60	2.41	4.49	0.19	1.75
65. 06 December 2017–W. Tower	06DecWT	13.3	37.3	2.29	0.28	0.19	2.05	2.08	0.18	4.73
66. 13 December 2017–W. Tower	13DecWT	12.5	37.8	3.87	0.37	0.10	0.54	3.20	0.49	17.28
67. 13 December 2017–Aretsou	13DecAR	12.4	37.4	5.47	0.45	0.23	3.54	2.04	1.09	16.33
68. 13 December 2017–Music Hall	13DecMH	12.8	37.7	6.42	0.32	0.11	1.28	2.68	0.13	0.93
69. 13 December 2017–Harbour	13DecHB	12.5	37.8	6.29	0.46	0.41	2.42	4.49	0.23	1.91
70. 19 December 2017–W. Tower	19DecWT	11.7	37.7	6.03	0.33	0.09	2.14	3.21	0.22	1.95
71. 27 December 2017–W. Tower	27DecWT	11.1	37.9	4.84	0.37	0.11	1.69	5.75	0.08	0.27
72. 03 January 2018–W. Tower	03JanWT	11.1	38	2.63	0.22	0.06	2.38	2.08	0.20	2.52
73. 10 January 2018–W. Tower	10JanWT	10.6	36.6	9.62	1.58	0.35	21.46	17.81	1.06	1.79
74. 10 January 2018–Aretsou	10JanAR	10.7	38.1	5.06	0.26	0.15	3.63	2.74	0.18	0.92
75. 10 January 2018–Music Hall	10JanMH	10.7	37.8	2.72	0.54	0.16	3.81	4.58	0.25	2.95
76. 10 January 2018–Harbour	10JanHB	10.6	37.2	3.43	0.55	0.16	7.67	7.27	1.12	8.31
77. 17 January 2018–W. Tower	17JanWT	10.4	37.8	1.46	0.26	0.04	1.01	2.24	0.12	1.02
78. 24 January 2018–W. Tower	24JanWT	9.6	37.9	2.28	0.29	0.08	2.35	2.06	0.18	1.15
79. 31 January 2018–W. Tower	31JanWT	10.7	38.2	2.48	0.28	0.07	3.00	2.24	0.20	1.15
80. 06 February 2018–W. Tower	06FebWT	11.9	38.4	2.27	0.32	0.09	2.78	3.22	0.28	2.82
81. 06 February 2018–Aretsou	06FebAR	11.8	38.1	3.14	0.21	0.07	2.80	5.93	0.34	2.91
82. 06 February 2018–Music Hall	06FebMH	11.7	38.4	1.89	0.17	0.06	0.44	1.91	0.26	2.30
83. 06 February 2018–Harbour	06FebHB	11.9	38	3.96	0.29	0.08	3.97	4.02	0.60	3.36

3.2. Plankton Diversity and Abundance

A total of 117 plankton morphospecies were identified in all four sampling sites during the study period (Supplementary Table S2). The taxonomic group of Bacillariophyta (diatoms) had the highest overall species richness as 44% of the total number of taxa belonged to this group, and was followed by Dinophyta (including also mixotrophic and heterotrophic dinoflagellates) (37% of the total number of taxa), Cryptophyta (5%), Haptophyta (3%), Chlorophyta (2%), Dictyochophyta (2%), and Euglenozoa (2%), while other groups (Cercozoa, Chrysophyceae, Telonemida, Xanthophyceae, and Ciliophora) contributed with < 2% of the total number of taxa (Supplementary Figure S3). In the four sampling sites, dinoflagellates were more diverse in terms of species richness during March 2017–November

2017, while diatoms appeared to be more diverse, in all sites, during December 2017–February 2018 (Supplementary Figure S3). The other taxonomic groups had a more or less consistent representation that altogether did not exceed in any case 40% of the total number of taxa in a sample.

The number of identified taxa varied among samples between 24 (22 March and 19 April in WT, and 9 May 2017 in MH) and 57 taxa on 8 November 2017 in WT (Supplementary Table S3), with the highest values in December–February when the measured water temperature was lower than 15 °C (Figure 2a). High variability was recorded in total cell abundance of phytoplankton reaching a maximum of 42,000 cells mL^{-1} on 19 July 2017 in WT, dominated by the diatom *Skeletonema costatum* (see Figure 2b). Heterotrophic dinoflagellates dominated by red tide forming *N. scintillans* exhibited highest values on 22 March 2017 in WT (>3250 cells mL^{-1}). The mean total taxa number and abundance were the only a-diversity estimators that were found significantly different in some paired comparisons between the different sites, based on t-tests; in particular: WT > AR, WT > MH and MH < HB (Supplementary Table S4; for a visualization of mean values of taxa number and abundance values in each sampling site, see Supplementary Figure S4). However, no significant differences in the distribution of the taxa relative abundances between sites were detected according to the Kolmogorov-Smirnov test ($p > 0.05$). The other a-diversity estimators calculated (Simpson, Shannon, Equitability, Evenness, and Berger–Parker), fluctuated during the study, and showed sometimes relatively low values, reflecting high dominance by one (or few) taxa, and high variation between taxa abundances within the community. In particular, the sampling dates with the low Simpson index (1-*D*) were: 22 March 2017 (0.10 in WT), 5 April 2017 (0.43 in WT), 31 May 2017 (0.22 in WT) and 18 October 2017 (0.29 in AR) (Supplementary Table S3).

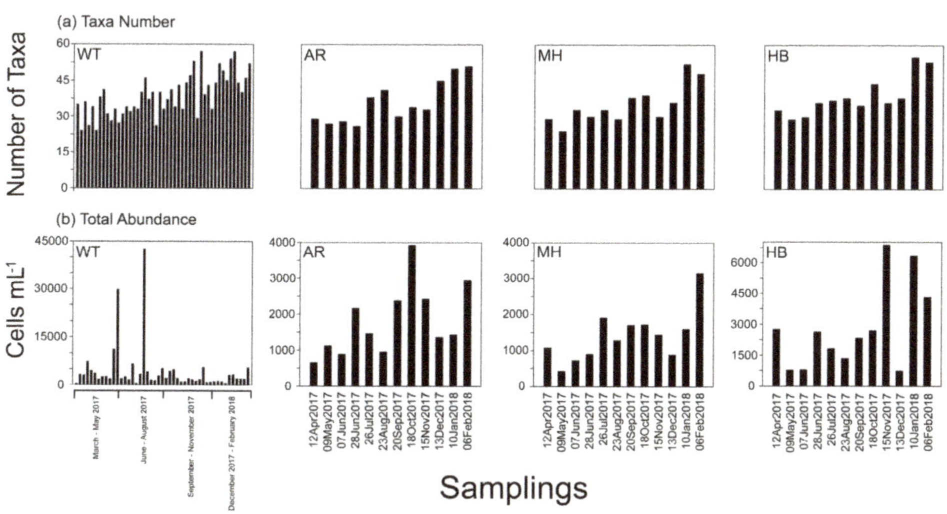

Figure 2. Number of taxa (**a**), and total abundance (**b**) in White Tower (WT), Aretsou (AR), Music Hall (MH), and Harbor (HB). For sampling codes, see also Table 2.

3.3. *Phytoplankton Blooms, Red Tides, and a Mucilage Aggregate Phenomenon*

Based on the plankton community composition and abundance during the study period, four major clusters were identified at a similarity level ~ 35% (Figure 3) grouping together the samplings irrespectively of the sample collection site, according to the sampling dates: March–June 2017 (Cluster I); July–October 2017 (Cluster II); November 2017 (Cluster III); and December 2017–February 2018 (Cluster IV). This is in accordance with the results of the t-test paired comparisons of a-diversity indices between sites that showed no significant differences in most occasions. The samples in each cluster were further grouped together based on higher similarity levels (>40% similarity). These groupings included a small number of samples taken in close dates and were characterized by phytoplankton blooms (>1000 cells mL^{-1}) of a taxonomic group or a single species, or/and red tides (Figure 3).

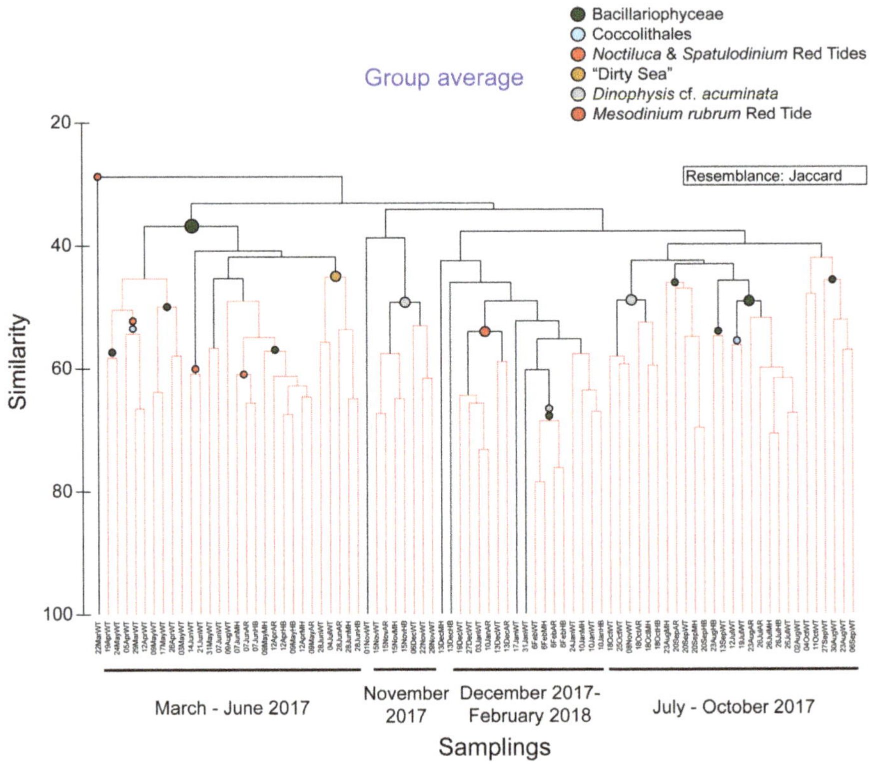

Figure 3. Cluster diagram according to Jaccard resemblance, calculated based on the non-transformed abundance (cells mL^{-1}) of taxa during the study. Red clades in the dendrogram indicate sections of the plot where the observed profile corresponds to similarities that are larger than those expected under null conditions (>99% of the confidence envelope), suggesting the presence of true structure within the data. The nodes represent the dominant taxa blooming during the period covered by the corresponding clades.

During the period March–June 2017 a persistent diatom bloom was detected at the White Tower (WT) site, due to the high abundances recorded throughout for the taxa of *Leptocylindrus danicus* (max: >7000 cells mL^{-1} on 24 May) and *Leptocylindrus minimus* (max abundance: >26000 cells mL^{-1} on 31 May). In specific samplings, i.e., on 24 May 2017, the taxon *S. costatum* additionally showed high concentrations reaching > 1000 cells mL^{-1}. The diatom bloom was accompanied by a Coccolithales bloom between 5 and 12 April 2017 (>5300 cells mL^{-1}).

On the other hand, conspicuous red tides, macroscopically visible, appeared in the front of the Bay at three occasions, during this period, making the water viscous. The red tides were detected at 22 March, 12 April, and 14–21 June 2017, and mainly consisted of the known red tide forming dinoflagellates *N. scintillans* and its close relative *Spatulodinium pseudonoctiluca*. Especially on 22 March 2017, the event was so intense that the sample consisted entirely of *N. scintillans* cells, reaching 3250 cells mL^{-1}, comprising > 99% of the total abundance. The co-occurrence of these species with bloom-forming, mucilage-producing diatoms, e.g., *Cylindrotheca closterium*, *Chaetoceros* spp., *L. minimus*, *L. danicus*, *S. costatum*, the haptophyte *Phaeocystis* sp. and the dinoflagellate *Gonyaulax* cf. *fragilis*, were observed. They were producing or being embedded in mucilage, before and during the development of an extreme aggregation of mucilage, between 28 June and 4 July 2017. *N. scintillans* was observed to feed on diatoms, and most commonly on *Chaetoceros* spp.

During the period July–October, diatoms remained in high numbers, dominated by the taxa *Chaetoceros* spp. (max: >6000 cells mL^{-1} on 19 July), and more rarely *C. closterium* (max: >1800 cells mL^{-1} on 20 September). On 19 July and 13 September 2017, the taxon *S. costatum*, additionally showed high abundances reaching > 25,000 and > 1500 cells mL^{-1}, respectively.

The diatom bloom was followed by an increase in the abundance of the harmful dinoflagellate *Dinophysis* cf. *acuminata*, during November 2017 (Figure 3). In particular, on 8 November 2017, *D.* cf. *acuminata* reached 120 cells mL^{-1}, in WT, while up to 350 cells mL^{-1} of this species were recorded at HB on the 15th November 2017.

The high abundances of *D.* cf. *acuminata* in November 2017 were followed by a red tide in all sampling sites during the period December 2017–February 2018. This bloom was dominated by the photosynthetic ciliate *Mesodinium rubrum*, first appearing in the Bay on 29 November 2017, and peaking from 13 December 2017 till 10 January 2018, reaching > 1000 cells mL^{-1} on 13 December. The period January–February 2018 was characterized by diatom dominance, i.e., the taxa *Chaetoceros tenuissimus* (max abundance > 2000 cells mL^{-1} on 6 February, WT), and *S. costatum* (reaching 1000 cells mL^{-1} on 3 January, WT).

Based on the IOC-UNESCO Taxonomic Reference List of Harmful Microalgae it can be stated that at least 11 out of 117 plankton taxa found in the present study have been reported as harmful. These taxa are the diatoms *Pseudonitzschia* cf. *delicatissima*, *Pseudonitzschia* cf. *multistriata*, *Pseudonitzschia* cf. *pseudodelicatissima*, and *Pseudonitzschia* cf. *pungens*, the dictyochophycean *Vicicitus globosus*, the haptophyte *Phaeocystis* sp. and the dinoflagellates *D.* cf. *acuminata*, *Dinophysis caudata*, *Karenia brevis*, *Karlodinium* spp., and the epiphytic *Prorocentrum* cf. *lima*. In particular, the diatoms *P.* cf. *delicatissima*, *P.* cf. *pseudodelicatissima*, and *P.* cf. *pungens* were detected in concentrations > 500 cells mL^{-1} at the White Tower (WT) site on 19 July 2017, just after the mucilage aggregate phenomenon. Relatively high abundances (270 cells mL^{-1}) were recorded for *K. brevis* at WT on the 28 June, right in the middle of the mucilage aggregate phenomenon. An extremely high bloom (>10,500 cells mL^{-1}) was observed at the same sampling point one year later (unpublished data).

3.4. Links of Environmental Parameters and Plankton Bloom, Red Tide, and Mucilaginous Aggregate-Forming Taxa

Connections between all detected taxa, including all phytoplankters and red tide/bloom-forming taxa, were investigated according to the MIC correlation coefficient. Only the strong connections between phytoplankters/red tide forming species and environmental parameters were visualized in network analysis (Figure 4). The strong connections represented MIC values corresponding to pre-calculated *p*-values (with $p < 0.01$), based on the total number of samples. Network analysis showed negative connections between salinity/water temperature and the majority of diatom taxa included in the network, and all Cryptophyta and Dictyochophyceae. However, the diatoms *Chaetoceros* spp., *S. costatum*, and *L. minimus*, all mucilage producers, were positively connected with salinity/water temperature. Dinophyta, on the other hand, showed mostly positive strong connections with salinity/water temperature and negative with POP (Figure 4). To note, the red tide forming *N. scintillans* showed strong positive connections with NH_4 and PO_4, while the other red tide forming species, *M. rubrum*, exhibited negative connections with salinity and water temperature. Finally, the harmful alga *D.* cf. *acuminata* displayed negative connection with water temperature (Figure 4).

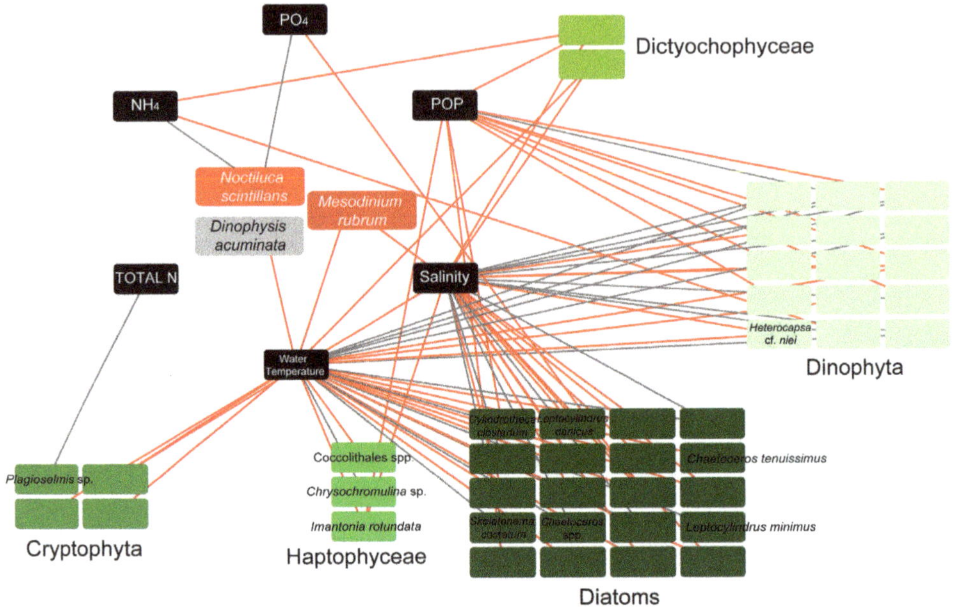

Figure 4. Network diagrams of highly significant connections (p-values < 0.01) based on the maximal information coefficient (MIC) scores between dominant taxa (comprising of > 10% of the total plankton abundance in at least one sample) and environmental parameters. Boxes (nodes) with indicated taxa names, represent bloom (detected with abundances > 1000 cells mL^{-1} in at least one sample throughout the samplings), and red tide forming taxa. To facilitate reading only bloom and red tide forming taxa are indicated. Black lines (edges) depict positive connections, and red edges depict negative connections.

Dinoflagellates were found to be significantly positively correlated with E.I. ($p < 0.001$, $r = 0.64$), N:P ($p < 0.01$, $r = 0.31$) and Si:P ($p < 0.05$, $r = 0.26$), while Cryptophyceae were significantly negatively correlated with Si:N ($p < 0.01$, $r = -0.31$) and Si:P ($p < 0.05$, $r = -0.22$).

4. Discussion

The reason for our focus on plankton community weekly dynamics of the urban marine system in the Thessaloniki Bay's front was motivated by the lack of relevant data in this particular system in eutrophication studies of Thermaikos Gulf, despite the recurrent phenomena of harmful algal blooms (HABs) and conspicuous mucilage phenomena. These phenomena are of great ecological importance for the coastal system and have significant socio-economic impact to the city's residents. After discussing nutrient pressure in the system, species diversity will be discussed, dominance of blooms and red tide forming species, and the key species which were the cause of the mucilage phenomenon verifying results of the marine eutrophication research and the related eutrophication symptoms [26].

4.1. Environmental Conditions

In the study area, a heavily modified marine water body according to WFD, annual mean salinity was 37.2 and close to the highest threshold value (37.5) for type IIA of the Mediterranean coastal water types that have been intercalibrated (applicable for phytoplankton) according to Commission Decision 2018/229/UE. This type of coastal water is considered moderately influenced by freshwater inputs, while the annual salinity average value is close to the boundary value (37.5) for type IIIE. Phytoplankton metrics have been intercalibrated only for type IIIE in Greece [35].

Throughout the study, nutrients (N and P) which are indication of eutrophication exhibited high values and were among the highest values reported for nutrient-rich coastal areas of the Mediterranean Sea [42–44]. In a recent comprehensive study on various coastal areas of Greece (1995–2007) influenced by anthropogenic activities, mainly by sewage and riverine outflows [45], the Thessaloniki Bay which is one of the most polluted coastal areas of Greece, exhibited the highest PO_4 (max 6.50 μmol L^{-1}) and NH_4 concentrations (max 15 μmol L^{-1}). Comparing to the highest values of nutrients during 1995–2007 [45], the values in this study (max 9.50 μmol L^{-1} and 160 μmol L^{-1} for PO_4 and NH_4, respectively) were even higher [45]. On the other hand, the highest NO_3 value (21.09 μmol L^{-1}) in the urban front of Thessaloniki Bay during our survey was slightly lower than the highest measured NO_3 value (23.5 μmol L^{-1}) during the period April–May 2012 [46]. Even so, both of them are extremely high for coastal sites, and are indicative of nitrogen pollution due to anthropogenic activities [47]. These outliers can be used as a sensitive tool for assessing water quality in coastal management studies according to Karydis [48], who showed that outliers are more sensitive in characterizing pollution/eutrophication levels than whole datasets, which usually include a large number of low values. The average N:P ratio in this study was higher (mean 25) compared to the N:P ratio (6.40) of the period 1995–2007 [45]. The high N:P ratio during the present study, in combination with the extreme NH_4 concentrations, may be linked to the relatively high contribution of dinoflagellates, such as *N. scintillans* and *S. pseudonoctiluca*, to plankton community biomass [49]. The much higher NH_4 values in 2017–2018 in combination with the high N:P ratio indicated nitrogen pollution, an increasing global problem [50]. Agriculture is the largest source of nitrogen pollution to many of the planet's coastal marine ecosystems [51].

In addition to high nutrient concentrations and ratios, the annual mean values of Chl *a* for all stations were higher than the values measured in the same area in 2012 [47]. Based on our data, ecological water quality can be classified as bad according to Simboura et al. [47]. Low Chl *a* values coincided both with low and high cell abundances in WT and HB. In most samples of WT the low Chl *a* coincided with high cell densities in spring and summer under high irradiance/day-length and temperature when *L. danicus* and *L. minimus* were the dominant species. These results may reflect the physiological state of these diatoms on their Chl *a* content due to the effect of temperature and irradiance [52,53]. Similarly, low Chl *a* was measured simultaneously with high *Leptocylindrus* densities in HB.

In addition to the evaluation of the eutrophication and the nitrogen pollution of the study area, based on the individual nutrient variations and their extreme values (outliers), the multimetric eutrophication index E.I. [30] is of great interest for coastal management. In our samples, the E.I. values were always > 0.83 (mean 2.56 after excluding outliers) indicating a heavily eutrophic system reflecting bad environmental status according to Pavlidou et al. [46]. The poor to bad water quality of Thessaloniki Bay according to the phytoplankton-based indices and the E.I. index used, is indicative of both nitrogen and phosphorus enrichment. There is evidence for Greek coastal waters that phytoplankton-based indices are highly sensitive to nitrogen enrichment while the E.I. index is highly sensitive to phosphorus enrichment [43]. It is noteworthy that according to Pavlidou et al. [46] the E.I. reflected the integral eutrophication status of a water body as a whole and has been proposed as a reliable tool regarding the assessment of eutrophication status, and the implementation of nutrient management strategies under the EU WFD and the EU MSFD.

4.2. Diversity and Composition of the Plankton Community

Various a-diversity indices have been used (Shannon, Simpson, Equitability, Evenness, Berger-Parker) to describe the structure of the community in terms of its species diversity, dominance and evenness. The species pool of the unicellular eukaryotic plankton community reflected by the a-diversity indices [54] was found similar in the four sampling sites, according to pairwise comparisons with t-test (see Table S4 for a-diversity pairwise comparisons). Additionally, the Kolmogorov–Smirnov test showed no significant differences on the distribution of taxa between sampling sites. A seed bank of the local pool, persistent or transient according to Partel et al. [54] and the plankton life history traits [55] contributed to maintain relatively high biodiversity in this urban degraded marine environment. Even though a-diversity indices showed no significant differences between sites, when considering only the mean number of species identified in this study, significantly lower mean values were found in HB, a site with the highest ammonia and nitrite nitrogen pollution (Table 2), and a generally stressed area because of the harbor daily activity. The lower species diversity might be explained by environmental change consisting of several stressors, which can cause stress-induced community sensitivity [56]. The impacts of environmental stress on biodiversity are well known [57].

Diatoms were the most diverse taxonomic group with the highest species numbers in all sites during the period of December 2017–February 2018 in contrast to dinoflagellates that were more diverse during the period of March 2017–November 2017 (Supplementary Figure S3). The different temporal pattern of diatom and dinoflagellate species richness, as also shown by their contrasting relationships to water temperature and salinity, might be explained by their different response to vertical water mixing versus stratification conditions [58]. However, very few diatoms are strictly restricted to the periods of deep mixing, while sinking is an important factor for the growth and bloom formation only for large, sinking diatoms and large, non-sinking dinoflagellates [55].

4.3. Phytoplankton Blooms

During the study, at least one taxon per sample was recorded in bloom abundances, dominating the plankton community. The taxa that were detected in bloom abundances mainly belonged to diatoms, mainly *Chaetoceros* spp., *C. closterium, L. minimus, L. danicus*, and *S. costatum*. The persistent growth and bloom formations by these diatoms under various turbulent and stratification conditions can be explained by their small cell size in combination with mucilage production [55,58,59].

These species with extended blooms were the most important constituents of Thermaikos Gulf phytoplankton 30 years ago, when untreated sewage entered the Gulf [18]. Even though no connections were found between them and nutrient concentrations according to network analysis in the present study, it is well known that under P-limited conditions, certain diatoms become increasingly dominant with increasing Si:P ratios [60]. The most persistent diatoms blooms during our survey coincided with the highest Si:P ratios (>20). Dense diatom blooms in marine ecosystems suffering from eutrophication can generate highly dominant diatom communities within phytoplankton assemblages [61]. Nevertheless, apart from the proved impacts of nutrient concentration and ratios on the occurrence of algal blooms [62], in many cases it seems that algal bloom proliferation is more complicated, and the quantity and the ratio of inorganic nutrients alone cannot sufficiently explain high abundance blooms of extended duration [8].

The known harmful species *D.* cf. *acuminata*, *P.* cf. *delicatissima*, *P.* cf. *multistriata*, *P.* cf. *pseudodelicatissima*, *P.* cf. *pungens*, *D. caudata*, *K. brevis*, *Karlodinium* spp., and the epiphytic *P.* cf. *lima*, with worldwide distribution, were detected in relatively high abundances, but not exceeding bloom densities during individual sampling dates. Furthermore, the known harmful alga *V. globosus* was recorded occasionally in live water samples taken during the sampling period, but its cells usually could not be preserved with Lugol's solution. These plankton species have been previously reported in the Thermaikos Gulf [18,21,63] indicating a persistent seed bank of the local pool [64]. Previous studies reported evidence for a diverse cyst bank, with high recorded abundance of cysts even in periods when the corresponding species were absent from the water column [65]. These cysts were associated with the formation of dense algal blooms in the water column and a high risk of HABs, as could be the case of the *Dinophysis* bloom observed in the present study during the period October–November 2017, and a short-term excessive *Karenia* bloom of extreme densities that was observed in spring 2018 (unpublished data). The urban Thessaloniki Bay exhibits the sustained increases in algal blooms and in HABs in accordance with high nutrient levels, similar to reports in other coastal areas of Mediterranean Sea and the Black Sea [23,66].

4.4. Red Tides

Several occasions of macroscopically visible red tides were documented over a temperature range of 10 to 25 °C, and a salinity range of 36 to 38.5. They were attributed to the known red tide forming dinoflagellates of *N. scintillans* together with its close taxonomic relative *S. pseudonoctiluca*, and the photosynthetic ciliate *M. rubrum*.

Noctiluca scintillans is one of the most important red tide forming dinoflagellate worldwide in the water temperature range of 10–25 °C, and salinity range of 28 to 36 in eutrophic areas dominated by diatoms [15,67], similar to our study area. *Noctiluca* red tides have been linked to eutrophication in several areas of the world and especially in the Black Sea, the Sea of Marmara [68–70], the Aegean Sea [71], and Adriatic Sea [72]. In contrast, *S. pseudonoctiluca* has been reported rarely, although new records from many areas suggest a cosmopolitan distribution. Its distribution has been underestimated due to its complex life cycle, morphological variability, and taxonomic issues in its identification [73]. In the Mediterranean Sea among these red tide forming species, another species of *Spatulodinium* has been found based on DNA analysis [73]. In the Mediterranean Sea, *Spatulodinium* showes a wide range of temperature preference, similar to the temperature range (19–30 °C) reported in the Mexican Pacific [74]. Both *N. scintillans* and *S. pseudonoctiluca* have been considered exclusively heterotrophic and inclusion of diatoms, dinoflagellates, and dictyochophytes have been observed in their cells [74].

The red tide forming *N. scintillans* terminated its growth in our study area by the increase of water temperature above 25 °C, as in many other studies, but temperature did not correlate with the start of its growth ([67] and references therein). A rich food supply of a broad spectrum of food items (from bacteria to fish eggs) is needed to start massive growth and formation of red tides, while availability of phytoplankton as a prey is a key factor [67,75]. Particularly, *Noctiluca* red tides are known to coincide or follow diatom blooms [16,67,76]. A strong temporal overlapping of *N. scintillans* and diatoms blooms has been also observed in the present study. High numbers of *N. scintillans* (>400 cells mL^{-1}) coincided with high numbers of *Chaetoceros* spp., *L. minimus*, *S. costatum*, and *C. closterium* cells. Different species of diatoms (mostly *Chaetoceros* spp.) have been observed in food vacuoles of *N. scintillans* in agreement with other studies [77,78]. Additionally, *N. scintillans* was feeding on harmful *Dinophysis* spp. *Noctiluca scintillans* containing toxigenic *Dinophysis* and *Pseudonitzschia* species may act as a vector of toxigenic algae to higher trophic levels or transport to shellfish aquaculture [79]. On the other hand, grazing pressure by *N. scintillans* on the growth of other toxigenic dinoflagellates should be considered as a potential regulator of phytoplankton toxins production [80]. This is of particularly interest in our study area, due to its close vicinity to the biggest mussel culture of Greece, where a harvest ban is often implemented due to *Dinophysis* spp. abundance > 1 cell mL^{-1} [81].

Accumulation of *N. scintillans* cells in the surface water, forming a red tide, was observed under calm weather days (generally with daily mean wind speed < 3 m s^{-1}) in this urban front of the bay protected from intense water circulation [82]. It is established that meteorological conditions and topography are crucial factors for red tide formation [75]. In Thessaloniki Bay, *N. scintillans* appeared to prefer higher salinity (>36) relative to those found in other studies [16,67,70]. Based on our results and *N. scintillans* abundance dynamics during spring-early summer in the Black Sea and the Northern Adriatic Sea, it is suggested that weather forecasts, and in particular wind speed projections, can be used for medium-term prediction of red tides [83].

A strong positive connection between *N. scintillans* cell abundance and NH_4 and PO_4 in our study area might indicate nutrient regeneration by this heterotrophic dinoflagellate and contribution to the local nutrient pool. The significant role of *N. scintillans* as a nutrient regenerator and an efficient recycler of nitrogen has been linked to extremely high concentrations of nitrogen in its cells and excretion regulated by nutrient quality of its food items. Nutrient liberation of senescent cells would stimulate the phytoplankton growth near the red tide patches while improving the food quality for *N. scintillans* [78]. NH_4 regeneration by *N. scintillans* in coastal seas has been reported by Montani et al. [84] whereas high ammonia concentrations released from *Noctiluca* cells during the decay process of the red tide were also shown by Schaumann et al. [85]. NH_4 also increases during decline bloom phase indicating release of intracellular NH_4 accumulated through *Noctiluca* grazing according to Baliarsingh et al. [86]. Direct toxicity to fish by ammonium/ammonia is possible although at seawater pH, approximately 5% of total ammonia is unionized NH_3 [87].

Mesodinium rubrum is a globally distributed photosynthetic ciliate that sometimes causes red tides in coastal waters [88]. *M. rubrum* is a marine plankton of great cytological, physiological, and evolutionary interest, which has an exceptional type of cellular organization not realized by other species, supported by organelle robbery [89]. *M. rubrum* and its accompanying cryptophytes showed a strong positive correlation ($p < 0.001$, $r = 0.59$), according to correlation analysis in the present study. *M. rubrum* reached high numbers (>700 cells mL^{-1}) in December (temperature range from 11.1 to 13.3 °C, and salinity range from 37.3 to 38.0), just after the drop of the *D.* cf. *acuminata* maxima at all sites. The red tide formed was spatially extended and the abundance of *M. rubrum* was found negatively correlated with both temperature and salinity. An important factor for *M. rubrum* seasonal dynamics and its short-lived bloom in the study area seems to be the persistent occurrence of *Dinophysis* spp. and several common heterotrophic dinoflagellates, which it is known to feed [90,91].

4.5. Mucilage Aggregates

Noctiluca red tides have been linked to mucilage phenomena, either as a shift from red tides to these events [92] or an overlapping that could be observed in Lapseki coastal area of the Dardanelles in early summer where gelatinous surface layers were recorded [16]. In Thessaloniki Bay, mucilage aggregates appeared on 22 June 2017, characterized by creamy whitish-brownish and gelatinous surface layers [93], which became progressively darker with age. Before the appearance of the phenomenon in Thessaloniki Bay, the plankton community consisted of known mucilage producing species such as the common diatoms in the bay *C. closterium, L. minimus, L. danicus, S. costatum*, the dinoflagellate *G.* cf. *fragilis* and the slime producing red tide dinoflagellates *N. scintillans*, and *S. pseudonoctiluca* and the foam forming *Phaeocystis* sp. Many studies have been published that relate diatom extracellular polymer production with the well-known phenomenon of marine mucilage in the Adriatic Sea, in particular the diatom species *C. closterium* [94]. This species has been regularly observed as dominant in the mucilage macroaggregates and it has been demonstrated experimentally that polysaccharides from it can form a gel network similar to the macroscopic gel phase occurring in the northern Adriatic Sea irrespective of any bacterial mediation or interaction with inorganic particles [95]. In our study, *C. closterium* has been observed abundant before, during and after the mucilage phenomenon, within abundant transparent freshly formed mucilage, whereas *Chaetoceros* spp. and *S. costatum* chains were also embedded in the mucilage. The dinoflagellate *G.* cf. *fragilis* was observed actively producing mucilage in the samples from June 2017 similarly as in the Emilia-Romagna coast (Northern Adriatic Sea) by Pompei et al. [96]. The same phytoplankton mucilage producers, i.e., *G.* cf. *fragilis, S. costatum*, and *C. closterium* were identified as abundant species also in a mucilage phenomenon in the Sea of Marmara [97]. The well-known foam-forming *Phaeocystis pouchetii* caused mucilage problems in the Evoikos Gulf [98].

Small mucilage aggregates were observed both by active and decaying *N. scintillans* cells during red tide formation and termination. Decaying *N. scintillans* cells contribute high amounts of organic matter to the local pool while active cells excrete mucus for trapping food items [78]. The aggregates formed by decaying *N. scintillans* sampled in the Northern Adriatic Sea presented a similar chemical - biochemical composition to the different typologies of mucilage aggregates in the same area [94]. This showed that the organic matter of *N. scintillans* could form a part of the mucilage organic matter in the Adriatic Sea. The accumulation of excess autochthonous organic material (dead and alive material) from the preceding red tides and phytoplankton blooms producing mucilage (from late March to June) and the mucilage they produce, in combination with the hydrodynamic conditions in the Bay (initiation of thermal stratification in May) [82] are suggested as main factors for the formation of the creamy and gelatinous surface layer in the urban Thessaloniki Bay. Our results agree with Umani et al. [10] study on the microbial community of a coastal area in the northern Adriatic Sea with frequent reports of mucilage aggregates, suggesting that mucilage is derived from accumulated slow-to-degrade dissolved organic matter. The months preceding the mucilage events (March–May) in the Northern Adriatic Sea were assumed to be an 'incubation' period. Mucilage was the consequence of a coupling between the accumulation of organic matter and the temporal pattern of meteorological and oceanographic conditions [99] similar to our observations. Strong north winds (>10 m s^{-1}) in the beginning of July 2017 were successful to degrade the gelatinous surface layer suddenly and disperse it as small mucilage aggregates. After a week, microscopic aggregates were observed concentrated above the pycnocline (5 m) in deeper areas in the Thermaikos Gulf [82].

5. Conclusions

During the study period, analysis of the weekly water samples from the urban coastal frontal zone of Thessaloniki Bay (Thermaikos Gulf) provided an outlook of the effects of eutrophication in this Mediterranean urban environment with further implications on marine eutrophication research and coastal management. In the majority of the samples, phytoplankton abundance, and nutrient concentrations indicated high eutrophic conditions and bad environmental status according to the implementation of the EU WFD and the EU MSFD. In addition, in all samples, the Eutrophication Index (E.I.) indicated a heavily eutrophic system, which was characterized by persistent phytoplankton blooms and conspicuous red tides. The phytoplankton blooms were dominated by the diatom species *Cylindrotheca closterium*, *Chaetoceros* spp., *Leptocylindrus minimus*, *Leptocylindrus danicus*, and *Skeletonema costatum* reaching high abundances during the spring-summer 2017, while the species *Chaetoceros tenuissimus* and *S. costatum* formed blooms during January–February 2018. Red tides of the species *Noctiluca scintillans* accompanied with *Spatulodinium pseudonoctiluca* in March 2017, and the species *Mesodinium rubrum* in December 2017 were observed in the Bay, while a mucilage aggregate phenomenon formed by the mucilage-producers *C. closterium*, *Chaetoceros* spp., *L. minimus*, *L. danicus*, *S. costatum*, *Phaeocystis* sp., and *Gonyaulax* cf. *fragilis* was observed in June 2017. These mucilage producers were linked to high temperatures/low salinity, while, on the other hand, red tide forming *N. scintillans* was linked to high nitrogen and phosphorus concentrations and higher salinities. These harmful events, along with the occurrence of several harmful algae, such as the known toxin-producer *Dinophysis* cf. *acuminata*, illustrate the need for continuous monitoring of target indicators of nutrient pollution, ecological water quality and environmental status in the Bay. In the prism of climate change and the increase of eutrophication conditions in coastal areas, this study sounds the alarm and highlights the need to reduce the causes contributing to the bad environmental status and the development of the described phenomena causing severe socio-economic impacts to the public.

Supplementary Materials: The following are available online at http://www.mdpi.com/1424-2818/11/8/136/s1; **Figure S1.** Environmental parameters examined in the four sampling sites; **Figure S2.** Annual mean values of the environmental parameters examined in the four sampling sites; **Figure S3.** Pie chart: Relative number of taxa belonging to major high-level unicellular eukaryotic taxonomic groups in all samplings; **Figure S4.** Bars showing the mean values of the number of taxa and abundance (cells mL^{-1}) in the four sampling sites; **Table S1.** Differences in physical and chemical parameters among the different sites; **Table S2.** Species list of unicellular eukaryotes found in Thessaloniki's Bay during the study period; **Table S3.** Species number, total cell abundance, and alpha diversity measurements per sample; **Table S4.** Differences in taxa number, cell abundance, and the diversity indices among the different sites,

Author Contributions: Conceptualization: S.G. and M.M.-G.; methodology: S.G., N.S. and M.M.-G.; validation: S.G., N.S. and M.M.-G.; formal analysis: S.G., and M.M.-G.; investigation: S.G. and N.S.; resources: U.S. and M.M.-G.; data curation: S.G. and N.S.; writing–original draft preparation: S.G., N.S., and M.M.-G.; writing–review and editing: S.G., U.S., and M.M.-G.; visualization: S.G.; supervision: M.M.-G.; funding acquisition: S.G., U.S., and M.M.-G.

Funding: This research is implemented through IKY scholarships programme and co-financed by the European Union (European Social Fund - ESF) and Greek national funds through the action entitled "Reinforcement of Postdoctoral Researchers", in the framework of the Operational Programme "Human Resources Development Program, Education and Lifelong Learning" of the National Strategic Reference Framework (NSRF) 2014–2020.

Conflicts of Interest: The authors declare no conflict of interest.

References

1. Rabalais, N.N.; Turner, R.E.; Díaz, R.J.; Justić, D. Global change and eutrophication of coastal waters. *ICES J. Mar. Sci.* **2009**, *66*, 1528–1537. [CrossRef]
2. Garnier, J.; Beusen, A.; Thieu, V.; Billen, G.; Bouwman, L. N:P:Si nutrient export ratios and ecological consequences in coastal seas evaluated by the ICEP approach. *Glob. Biogeochem. Cycl.* **2010**, *24*. [CrossRef]
3. Kroeze, C.; Hofstra, N.; Ivens, W.; Löhr, A.; Strokal, M.; van Wijnen, J. The links between global carbon, water and nutrient cycles in an urbanizing world-the case of coastal eutrophication. *Curr. Opin. Environ. Sustain.* **2013**, *55*, 566–572. [CrossRef]
4. Cardona, Y.; Bracco, A.; Villareal, T.A.; Subramaniam, A.; Weber, S.C.; Montoya, J.P. Highly variable nutrient concentrations in the Northern Gulf of Mexico. *Deep-Sea Res. Pt. II* **2016**, *129*, 20–30. [CrossRef]
5. Smith, V.H. Eutrophication of freshwater and coastal marine ecosystems: A global problem. *Environ. Sci. Pollut. Res.* **2003**, *10*, 126–139. [CrossRef]
6. Turkoglu, M.; Erdogan, Y. Diurnal variations of summer phytoplankton and interactions with some physicochemical characteristics under eutrophication of surface water in the Dardanelles (Çanakkale Strait, Turkey). *Turk. J. Biol.* **2010**, *34*, 211–225.
7. Turkoglu, M. Temporal variations of surface phytoplankton, nutrients and chlorophyll-a in the Dardanelles (Turkish Straits System): A coastal station sample in weekly time intervals. *Turk. J. Biol.* **2010**, *34*, 319–333.
8. Heisler, J.; Glibert, P.M.; Burkholder, J.M.; Anderson, D.M.; Cochlan, W.; Dennison, W.C.; Gobler, C.; Dortch, Q.; Heil, C.; Humphries, E.; et al. Eutrophication and harmful algal blooms: A scientific consensus. *Harmful Algae* **2008**, *8*, 3–13. [CrossRef] [PubMed]
9. Anderson, D.M.; Cembella, A.D.; Hallegraeff, G.M. Progress in understanding harmful algal blooms: Paradigm shifts and new technologies for research, monitoring, and management. *Annu. Rev. Mar. Sci.* **2012**, *4*, 143–176. [CrossRef]
10. Umani, S.F.; Del Negro, P.; Larato, C.; De Vittor, C.; Cabrini, M.; Celio, M.; Bingol, K.; Falconi, C.; Tamberlich, F.; Azam, F. Major inter-annual variations in microbial dynamics in the Gulf of Trieste (northern Adriatic Sea) and their ecosystem implications. *Aquat. Microb. Ecol.* **2007**, *46*, 163–175. [CrossRef]
11. Cibic, T.; Cerino, F.; Karuza, A.; Fornasaro, D.; Comici, C.; Cabrini, M. Structural and functional response of phytoplankton to reduced river inputs and anomalous physical-chemical conditions in the Gulf of Trieste (northern Adriatic Sea). *Sci. Total Environ.* **2018**, *636*, 838–853. [CrossRef] [PubMed]
12. Parsons, M.L.; Dortch, Q. Sedimentological evidence of an increase in *Pseudo-nitzschia* (Bacillariophyceae) abundance in response to coastal eutrophication. *Limnol. Oceanogr.* **2002**, *47*, 551–558. [CrossRef]
13. Li, J.; Gilbert, P.M.; Zhou, M. Temporal and spatial variability in nitrogen uptake kinetics during harmful dinoflagellate blooms in the East China Sea. *Harmful Algae* **2010**, *9*, 531–539. [CrossRef]
14. Dela-Cruz, J.; Middleton, J.H.; Suthers, I.M. The influence of upwelling, coastal nutrients and water temperature on the distribution of the red tide dinoflagellate, *Noctiluca scintillans*, along the east coast of Australia. *Hydrobiologia* **2008**, *598*, 59–75. [CrossRef]
15. Harrison, P.J.; Furuya, K.; Glibert, P.M.; Xu, J.; Liu, H.B.; Yin, K.; Lee, J.H.W.; Anderson, D.M.; Gowen, R.; Al-Azri, A.R.; et al. Geographical distribution of red and green *Noctiluca scintillans*. *Chin. J. Oceanol. Limnol.* **2011**, *29*, 807–831. [CrossRef]
16. Turkoglu, M. Red tides of the dinoflagellate *Noctiluca scintillans* associated with eutrophication in the Sea of Marmara (the Dardanelles, Turkey). *Oceanologia* **2013**, *55*, 709–732. [CrossRef]
17. Krestenitis, Y.N.; Kombiadou, K.D.; Androulidakis, Y.S. Interannual variability of the physical characteristics of North Thermaikos Gulf (NW Aegean Sea). *J. Mar. Syst.* **2012**, *96*, 132–151. [CrossRef]
18. Nikolaidis, G.; Moustaka-Gouni, M. The structure and dynamics of phytoplankton assemblages from the inner part of the Thermaikos Gulf, Greece. I. Phytoplankton composition and biomass from May 1988 to April 1989. *Hegol. Mar. Res.* **1990**, *44*, 487. [CrossRef]
19. Karageorgis, A.P.; Skourtos, M.S.; Kapsimalis, V.; Kontogianni, A.D.; Skoulikidis, N.T.; Pagou, K.; Nikolaidis, N.P.; Drakopoulou, P.; Zanou, B.; Karamanos, H.; et al. An integrated approach to watershed management within the DPSIR framework: Axios River catchment and Thermaikos Gulf. *Reg. Environ. Chang.* **2005**, *5*, 138–160. [CrossRef]
20. Mihalatou, H.M.; Moustaka-Gouni, M. Pico-, nano-, microplankton abundance and primary productivity in a eutrophic coastal area of the Aegean Sea, Mediterranean. *Int. Rev. Hydrobiol.* **2002**, *87*, 439–456. [CrossRef]

21. Genitsaris, S.; Kormas, K.A.; Moustaka-Gouni, M. Airborne algae and cyanobacteria: Occurrence and related health effects. *Front. Biosci.* **2011**, *3*, 772–787.
22. Friligos, N.; Kondylakis, J.C.; Psyllidou-Giouranovits, R.; Georgakopoulou-Gregoriadou, E. Eutrophication and phytoplankton abundance in the Thermaikos Gulf, Greece. *Fresenius Environ. Bull.* **1997**, *6*, 27–31.
23. Moncheva, S.; Gotsis-Skretas, O.; Pagou, K.; Krastev, A. Phytoplankton blooms in Black Sea and Mediterranean coastal ecosystems subjected to anthropogenic eutrophication: Similarities and differences. *Estuar. Coastal Shelf Sci.* **2001**, *53*, 281–295. [CrossRef]
24. Koukaras, K.; Nikolaidis, G. Dinophysis blooms in Greek coastal waters (Thermaikos Gulf, NW Aegean Sea). *J. Plankton Res.* **2004**, *26*, 445–457. [CrossRef]
25. Reizopoulou, S.; Strogyloudi, E.; Giannakourou, A.; Pagou, K.; Hatzianestis, I.; Pyrgaki, C.; Granéli, E. Okadaic acid accumulation in macrofilter feeders subjected to natural blooms of *Dinophysis acuminata*. *Harmful Algae* **2008**, *7*, 228–234. [CrossRef]
26. Ferreira, J.G.; Andersen, J.H.; Borja, A.; Bricker, S.B.; Camp, J.; Cardoso da Silva, M.; Garcés, E.; Heiskanen, A.S.; Humborg, C.; Ignatiades, L.; et al. Overview of eutrophication indicators to assess environmental status within the European Marine Strategy Framework Directive. *Estuar. Coast. Shelf Sci.* **2011**, *93*, 117–131. [CrossRef]
27. Weyl, P.K. On the change in electrical conductance of seawater with temperature. *Limnol. Oceanogr.* **1964**, *9*, 75–78. [CrossRef]
28. Jeffrey, S.W.; Humphrey, G.F. New spectrophotometric equation for determining chlorophyll a, b, c1 and c2. Biochem. *Physiol. Pflanz.* **1975**, *167*, 194–204.
29. Hansen, H.P.; Koroleff, F. Determination of nutrients. In *Methods of Seawater Analysis*, 3rd ed.; Grasshoff, K., Kremling, K., Ehrhardt, M., Eds.; Wiley VCH: Weinheim, Germany, 1999; pp. 159–228.
30. Primpas, I.; Tsirtsis, G.; Karydis, M.; Kokkoris, G.D. Principal component analysis: Development of a multivariate index for assessing eutrophication according to the European water framework directive. *Ecol. Indic.* **2010**, *10*, 178–183. [CrossRef]
31. Hasle, G.R.; Syvertsen, E.E.; Steidinger, K.A.; Tangen, K.; Tomas, C.R. *Identifying Marine Diatoms and Dinoflagellates*; Academic Press: Cambridge, MA, USA, 1996; p. 613.
32. Hasle, G.R.; Syvertsen, E.E.; Throndsen, J.; Steidinger, K.A.; Tangen, K.; Heimdal, B.R. *Identifying Marine Phytoplankton*; Academic Press: Cambridge, MA, USA, 1997; p. 858.
33. Gómez, F.; Furuya, K. *Kofoidinium*, *Spatulodinium* and other Kofoidiniaceans (Noctilucales, Dinophyceae) in the Pacific Ocean. *Eur. J. Protistol.* **2007**, *43*, 115–124.
34. Utermöhl, H. Zur Vervollkommnung der quantitativen Phytoplankton-Methodik. Mitt. *Int. Ver. Theor. Angew. Limnol.* **1958**, *9*, 1–38.
35. Integrated Monitoring and Assessment Programme of the Mediterranean Sea and Coast and Related Assessment Criteria. Available online: https://wedocs.unep.org/handle/20.500.11822/10947 (accessed on 12 August 2019).
36. Hammer, Ø.; Harper, D.A.T.; Ryan, P.D. PAST: Paleontological statistics software package for education and data analysis. *Paleontol. Electron.* **2001**, *4*, 9.
37. Morris, E.K.; Caruso, T.; Buscot, F.; Fischer, M.; Hancock, C.; Maier, T.S.; Meiners, T.; Müller, C.; Obermaier, E.; Prati, D.; et al. Choosing and using diversity indices: Insights for ecological applications from the German Biodiversity Exploratories. *Ecol. Evol.* **2014**, *4*, 3514–3524. [CrossRef] [PubMed]
38. Hillebrand, H.; Bennett, D.M.; Cadotte, M.W. Consequences of dominance: A review of evenness effects on local and regional ecosystem processes. *Ecology* **2008**, *89*, 1510–1520. [CrossRef] [PubMed]
39. Primer v6: User Manual/Tutorial. PRIMER-E Plymouth. Available online: https://www.researchgate.net/publication/285668711_PRIMER_v6_user_manualtutorial_PRIMER-E_Plymouth (accessed on 31 January 2006).
40. Reshef, D.N.; Reshef, Y.A.; Finucane, H.K.; Grossman, S.R.; McVean, G.; Turnbaugh, P.J.; Lander, E.S.; Mitzenmacher, M.; Sabeti, P.C. Detecting novel associations in large data sets. *Science* **2011**, *334*, 1518–1524. [CrossRef] [PubMed]
41. Smoot, M.E.; Ono, K.; Ruscheinski, J.; Wang, P.-L.; Ideker, T. Cytoscape 2.8: New features for data integration and network visualization. *Bioinformatics* **2011**, *27*, 431–432. [CrossRef] [PubMed]
42. Kontas, A.; Kucuksezgin, F.; Altay, O.; Uluturhan, E. Monitoring of eutrophication and nutrient limitation in the Izmir Bay (Turkey) before and after Wastewater Treatment Plant. *Environ. Int.* **2004**, *29*, 1057–1062. [CrossRef]

43. Simboura, N.; Pavlidou, A.; Bald, J.; Tsapakis, M.; Pagou, K.; Zeri, C.; Androni, A.; Panayotidis, P. Response of ecological indices to nutrient and chemical contaminant stress factors in Eastern Mediterranean coastal waters. *Ecol. Indic.* **2016**, *70*, 89–105. [CrossRef]
44. Tugrul, S.; Ozhan, K.; Akcay, I. Assessment of trophic status of the northeastern Mediterranean coastal waters: Eutrophication classification tools revisited. *Environ. Sci. Pollut. Res. Int.* **2018**, *26*, 14742–14754. [CrossRef]
45. Pavlidou, A. Nutrient distribution in selected coastal areas of Aegean Sea (East Mediterranean Sea). *J. Environ.* **2012**, *1*, 78–88.
46. Pavlidou, A.; Simboura, N.; Rousselaki, E.; Tsapakis, M.; Pagou, K.; Drakopoulou, P.; Assimakopoulou, G.; Kontoyiannis, H.; Panayotidis, P. Methods of eutrophication assessment in the context of the water framework directive: Examples from the Eastern Mediterranean coastal areas. *Cont. Shelf Res.* **2015**, *108*, 156–168. [CrossRef]
47. Simboura, N.; Tsapakis, M.; Pavlidou, A.; Asimakopoulou, G.; Pagou, K.; Kontoyiannis, H.; Zeri, C.; Krasakopoulou, E.; Rousselaki, E.; Katsiaras, N.; et al. Assessment of the environmental status in the Hellenic coastal waters (Eastern Mediterranean): from the Water Framework Directive to the Marine Strategy Framework Directive. *Mediterr. Mar. Sci.* **2015**, *16*, 46–64. [CrossRef]
48. Karydis, M. Environmental quality assessment based on the analysis of extreme values: A practical approach for evaluating eutrophication. *J. Environ. Sci. Heal.* **1994**, *29*, 775–791. [CrossRef]
49. Xiao, W.; Liu, X.; Irwin, A.J.; Laws, E.A.; Wang, L.; Chen, B.; Zeng, Y.; Huang, B. Warming and eutrophication combine to restructure diatoms and dinoflagellates. *Water Res.* **2018**, *128*, 206–216. [CrossRef] [PubMed]
50. Howarth, R.; Paerl, H.W. Coastal marine eutrophication: Control of both nitrogen and phosphorus is necessary. *Proc. Natl. Acad. Sci. USA* **2008**, *105*, E103. [CrossRef] [PubMed]
51. Howarth, R.W. Coastal nitrogen pollution: A review of sources and trends globally and regionally. *Harmful Algae* **2008**, *8*, 14–20. [CrossRef]
52. Verity, P.G. Effects of temperature, irradiance, and daylength on the marine diatom *Leptocylindrus danicus* Cleve. II. excretion. *J. Exp. Mar. Biol. Ecol.* **1981**, *55*, 159–169. [CrossRef]
53. Geider, R.J. Light and temperature dependence of the carbon to chlorophyll a ratio in microalgae and cyanobacteria: Implications for physiology and growth of phytoplankton. *New Phytol.* **1987**, *106*, 1–34. [CrossRef]
54. Partel, M.; Zobel, M.; Zobel, K.; Van Der Maarel, E. The species pool and its relation to species richness: Evidence from Estonian plant communities. *Oikos* **1996**, *75*, 111–117. [CrossRef]
55. Sommer, U.; Charalampous, E.; Genitsaris, S.; Moustaka-Gouni, M. Benefits, costs and taxonomic distribution of marine phytoplankton body size. *J. Plankton Res.* **2017**, *39*, 494–508. [CrossRef]
56. Stefanidou, N.; Genitsaris, S.; Lopez-Bautista, J.; Sommer, U.; Moustaka-Gouni, M. Effects of heat shock and salinity changes on coastal Mediterranean phytoplankton in a mesocosm experiment. *Mar. Biol.* **2018**, *165*, 154. [CrossRef]
57. Hillebrand, H.; Matthiessen, B. Biodiversity in a complex world: Consolidation and progress in functional biodiversity research. *Ecol. Lett.* **2009**, *12*, 1405–1419. [CrossRef] [PubMed]
58. Margalef, R. Turbulence and marine life. *Sci. Mar.* **1997**, *61*, 109–123.
59. Passow, U.; Alldredge, A.L. Aggregation of a diatom bloom in a mesocosm: the role of transparent exopolymer particles (TEP). *Deep-Sea Res. Pt. II* **1995**, *42*, 99–109. [CrossRef]
60. Sommer, U. Nutrient competition experiments with periphyton from the Baltic Sea. *Mar. Ecol. Prog. Ser.* **1996**, *140*, 161–167. [CrossRef]
61. Le Moal, M.; Gascuel-Odoux, C.; Ménesguen, A.; Souchon, Y.; Étrillard, C.; Levain, A.; Moatar, F.; Pannard, A.; Souchu, P.; Lefebvre, A.; et al. Eutrophication: A new wine in an old bottle? *Sci. Total Environ.* **2019**, *651*, 1–11. [CrossRef] [PubMed]
62. Smayda, T.J.; Reynolds, C.S. Strategies of marine dinoflagellate survival and some rules of assembly. *J. Sea Res.* **2003**, *49*, 95–106. [CrossRef]
63. Nikolaidis, G.; Koukaras, K.; Aligizaki, K.; Heracleous, A.; Kalopesa, E.; Moschandreou, K.; Tsolaki, E.; Mantoudis, A. Harmful microalgal episodes in Greek coastal waters. *J. Biol. Res.* **2005**, *3*, 77–85.
64. Spatharis, S.; Dolopsakis, N.P.; Economou-Amilli, A.; Tsirtsis, G.; Danielidis, D.B. Dynamics of potentially harmful microalgae in a confined Mediterranean Gulf-Assessing the risk of bloom formation. *Harmful Algae* **2009**, *8*, 736–743. [CrossRef]

65. Giannakourou, A.; Orlova, T.Y.; Assimakopoulou, G.; Pagou, K. Dinoflagellate cysts in recent marine sediments from Thermaikos Gulf, Greece: Effects of resuspension events on vertical cyst distribution. *Cont. Shelf Res.* **2005**, *25*, 2585–2596. [CrossRef]
66. Bodenau, N. Microbial blooms in the Romanian area of the Black Sea and contemporary eutrophication conditions. In *Toxic Phytoplankton Blooms in the Sea*; Smayda, T.J., Shimizu, Y., Eds.; Elsevier: Amsterdam, The Netherland, 1993; pp. 203–209.
67. Tsai, S.-F.; Wu, L.-Y.; Chou, W.-C.; Chiang, K.-P. The dynamics of a dominant dinoflagellate, *Noctiluca scintillans*, in the subtropical coastal waters of the Matsu archipelago. *Mar. Pollut. Bull.* **2018**, *127*, 553–558. [CrossRef] [PubMed]
68. Turkoglu, M.; Koray, T. Phytoplankton species succession and nutrients in the Southern Black Sea (Bay of Sinop). *Turk. J. Bot.* **2002**, *26*, 235–252.
69. Turkoglu, M.; Koray, T. Algal blooms in surface waters of the Sinop Bay in the Black Sea, Turkey. *Pak. J. Biol. Sci.* **2004**, *7*, 1577–1585.
70. Kopuz, U.; Feyzioglu, A.M.; Valente, A. An unusual red-tide event of *Noctiluca scintillans* (Macartney) in the southeastern Black Sea. *Turk. J. Fish. Aquat. Sci.* **2014**, *14*, 261–268. [CrossRef]
71. Koray, T.; Buyukisik, B.; Parlak, H.; Gokpinar, S. Final reports on research projects dealing with eutrophication and heavy metal accumulation. In *Eutrophication Processes and Algal Blooms (Red-Tides) in Izmir Bay*; UNEP: Athens, Greece, 1996; Volume 104, pp. 1–26.
72. Umani, S.F.; Beran, A.; Parlato, S.; Virgilio, D.; Zollet, T.; De Olazabal, A.; Lazzarini, B.; Cabrini, M. *Noctiluca scintillans* Macartney in the Northern Adriatic Sea: Long-term dynamics, relationships with temperature and eutrophication, and role in the food web. *J. Plankton Res.* **2004**, *26*, 545–561. [CrossRef]
73. Gómez, F. Diversity and distribution of noctilucoid dinoflagellates (Noctilucales, Dinophyceae) in the open Mediterranean Sea. *Acta Protozool.* **2010**, *49*, 365–372.
74. Gárate-Lizárraga, I.; García-Domínguez, F.; Pérez-Cruz, B.; Díaz-Ortiz, J.A. First record of *Cochlodinium convoltum* and *C. helicoides* (Gymnodiniales: Dinophyceae) in the Gulf of California. *Rev. Biol. Mar. Oceanog.* **2011**, *46*, 495–498. [CrossRef]
75. Miyaguchi, H.; Fujiki, T.; Kikuchi, T.; Kuwahara, V.S.; Toda, T. Relationship between the bloom of *Noctiluca scintillans* and environmental factors in the coastal waters of Sagami Bay, Japan. *J. Plankton Res.* **2006**, *28*, 313–324. [CrossRef]
76. Dela-Cruz, J.; Ajani, P.; Lee, R.; Pritchard, T.; Suthers, I. Temporal abundance patterns of the red tide dinoflagellate *Noctiluca scintillans* along the southeast coast of Australia. *Mar. Ecol. Prog. Ser.* **2002**, *236*, 75–88. [CrossRef]
77. Tiselius, P.; Kiørboe, T. Colonization of diatom aggregates by the dinoflagellate *Noctiluca scintillans*. *Limnol. Oceanogr.* **1998**, *43*, 154–159. [CrossRef]
78. Zhang, S.; Liu, H.; Ke, Y.; Li, B. Effect of the silica content of diatoms on protozoan grazing. *Front. Mar. Sci.* **2017**, *4*, 202. [CrossRef]
79. Escalera, L.; Pazos, Y.; Morono, A.; Reguera, B. *Noctiluca scintillans* may act as a vector of toxigenic microalgae. *Harmful Algae* **2007**, *6*, 317–320. [CrossRef]
80. Frangópulos, M.; Spyrakos, E.; Guisande, C. Ingestion and clearance rates of the red *Noctiluca scintillans* fed on the toxic dinoflagellate *Alexandrium minutum* (Halim). *Harmful Algae* **2011**, *10*, 304–309. [CrossRef]
81. Vlamis, A.; Katikou, P. Climate influence on *Dinophysis* spp. spatial and temporal distributions in Greek coastal water. *Plankton Benthos Res.* **2014**, *9*, 15–31. [CrossRef]
82. Monitoring the Marine Environment of Thermaikos Gulf. Available online: https://www.researchgate.net/publication/326292943_Monitoring_the_marine_environment_of_Thermaikos_Gulf/link/5b4441c40f7e9bb59b1b265a/download (accessed on 10 July 2019).
83. Mikaelyan, A.S.; Malej, A.; Shiganova, T.A.; Turk, V.; Sivkovitch, A.E.; Musaeva, E.I.; Kogovšek, T.; Lukasheva, T.A. Populations of the red tide forming dinoflagellate *Noctiluca scintillans* (Macartney): A comparison between the Black Sea and the northern Adriatic Sea. *Harmful Algae* **2014**, *33*, 29–40. [CrossRef]
84. Montani, S.; Pithakpol, S.; Tada, K. Nutrient regeneration in coastal seas by *Noctiluca scintillans*, a red tide-causing dinoflagellate. *J. Mar. Biotechnol.* **1998**, *6*, 224–228.
85. Schaumann, K.; Gerdes, D.; Hesse, K. Hydrographic and biological characteristics of a *Noctiluca scintillans* red tide in the German Bight, 1984. *Meeresforschung* **1988**, *32*, 77–91.

86. Baliarsingh, S.K.; Lotliker, A.A.; Trainer, V.L.; Wells, M.L.; Parida, C.; Sahu, B.K.; Srichandan, S.; Sahu, K.C.; Kumar, T.S. Environmental dynamics of red *Noctiluca scintillans* bloom in tropical coastal waters. *Mar. Pollut. Bull.* **2016**, *111*, 277–286. [CrossRef]
87. Millero, F.J.; Graham, T.B.; Huang, F.; Bustos-Serrano, H.; Pierrot, D. Dissociation constants of carbonic acid in seawater as a function of salinity and temperature. *Mar. Chem.* **2006**, *100*, 80–94. [CrossRef]
88. Kang, N.S.; Lee, K.H.; Jeong, H.J.; Yoo, Y.D.; Seong, K.A.; Potvin, E.; Hwanga, Y.J.; Yoond, E.Y. Red tides in Shiwha Bay, western Korea: A huge dike and tidal power plant established in a semi-enclosed embayment system. *Harmful Algae* **2013**, *30*, S114–S130. [CrossRef]
89. Gustafson, D.E., Jr.; Stoesker, D.K.; Johnson, M.D.; Van Heukelem, W.F.; Sneider, K. Cryptophte algae are robbed of their organelles by the marine *Mesodinium rubrum*. *Nature* **2000**, *405*, 1049–1052. [CrossRef] [PubMed]
90. Park, M.G.; Kim, S.; Kim, H.S.; Myung, G.; Kang, Y.G.; Yih, W. First successful culture of the marine dinoflagellate *Dinophysis acuminata*. *Aquat. Microb. Ecol.* **2006**, *45*, 101–106. [CrossRef]
91. Lee, K.H.; Jeong, H.J.; Yoon, E.Y.; Jang, S.H.; Kim, H.S.; Yih, W. Feeding by common heterotrophic dinoflagellates and a ciliate on the red-tide ciliate *Mesodinium rubrum*. *Algae* **2014**, *29*, 153–163. [CrossRef]
92. Degobbis, D.; Umani, S.F.; Franco, P.; Malej, A.; Precali, R.; Smodlaka, N. Changes in the northern Adriatic ecosystem and the hypertrophic appearance of gelatinous aggregates. *Sci. Total Environ.* **1995**, *165*, 43–58. [CrossRef]
93. Precali, R.; Giani, M.; Marini, M.; Grilli, F.; Ferrari, C.R.; Pečar, O.; Paschini, E. Mucilaginous aggregates in the northern Adriatic in the period 1999–2002: Typology and distribution. *Sci. Total Environ.* **2005**, *353*, 10–23. [CrossRef] [PubMed]
94. Giani, M.; Berto, D.; Zangrando, V.; Castelli, S.; Sist, P.; Urbani, R. Chemical characterization of different typologies of mucilaginous aggregates in the Northern Adriatic Sea. *Sci. Total Environ.* **2005**, *353*, 232–246. [CrossRef]
95. Svetličić, V.; Žutić, V.; Radić, T.M.; Pletikapić, G.; Zimmermann, A.H.; Urbani, R. Polymer networks produced by marine diatoms in the Northern Adriatic Sea. *Mar. Drugs* **2011**, *9*, 666–679. [CrossRef]
96. Pompei, M.; Mazziotti, C.; Guerrini, F.; Cangini, M.; Pigozzi, S.; Benzi, M.; Palamidesi, S.; Boni, L.; Pistocchi, R. Correlation between the presence of *Gonyaulax fragilis* (Dinophyceae) and the mucilage phenomena of the Emilia-Romagna coast (northern Adriatic Sea). *Harmful Algae* **2003**, *2*, 301–316. [CrossRef]
97. Tüfekçi, V.; Balkis, N.; Beken, C.P.; Ediger, D.; Mantikçi, M. Phytoplankton composition and environmental conditions of a mucilage event in the Sea of Marmara. *Turk. J. Biol.* **2010**, *34*, 199–210.
98. Ignatiades, L.; Gotsis-Skretas, O. A review on toxic and harmful algae in Greek coastal waters (E. Mediterranean Sea). *Toxins* **2010**, *2*, 1019–1037. [CrossRef]
99. De Lazzari, A.; Berto, D.; Cassin, D.; Boldrin, A.; Giani, M. Influence of winds and oceanographic conditions on the mucilage aggregation in the Northern Adriatic Sea in 2003–2006. *Mar. Ecol.* **2008**, *29*, 469–482. [CrossRef]

© 2019 by the authors. Licensee MDPI, Basel, Switzerland. This article is an open access article distributed under the terms and conditions of the Creative Commons Attribution (CC BY) license (http://creativecommons.org/licenses/by/4.0/).

diversity

MDPI

Article

Photosynthetic Picoeukaryotes Diversity in the Underlying Ice Waters of the White Sea, Russia

Tatiana A. Belevich [1,2,*], Ludmila V. Ilyash [1], Irina A. Milyutina [2], Maria D. Logacheva [2] and Aleksey V. Troitsky [2,*]

1 Biological Faculty, Lomonosov Moscow State University, 119234 Moscow, Russia; ilyash@gmail.ru
2 Belozersky Institute of Physico-Chemical Biology, Lomonosov Moscow State University, 119992 Moscow, Russia; iramilyutina@yandex.ru (I.A.M.); maria.log@gmail.com (M.D.L.)
* Correspondence: belevich@mail.bio.msu.ru (T.A.B.); bobr@belozersky.msu.ru (A.V.T.)

Received: 5 February 2020; Accepted: 3 March 2020; Published: 5 March 2020

Abstract: The White Sea is a unique basin combining features of temperate and arctic seas. The current state of its biocenoses can serve as a reference point in assessing the expected desalination of the ocean as a result of climate change. A metagenomic study of under-ice ice photosynthetic picoeukaryotes (PPEs) was undertaken by Illumina high-throughput sequencing of the 18S rDNA V4 region from probes collected in March 2013 and 2014. The PPE biomass in samples was 0.03–0.17 μg C·L^{-1} and their abundance varied from 10 cells·mL^{-1} to 140 cells·mL^{-1}. There were representatives of 16 algae genera from seven classes and three supergroups, but Chlorophyta, especially Mamiellophyceae, dominated. The most represented genera were *Micromonas* and *Mantoniella*. For the first time, the predominance of Mantoniella (in four samples) and Bolidophyceae (in one sample) was observed in under-ice water. It can be assumed that a change in environmental conditions will lead to a considerable change in the structure of arctic PPE communities.

Keywords: White Sea; under-ice water; picoeukaryotes; *Micromonas*; *Mantoniella*; high-throughput sequencing; metagenomics; 18S rDNA

---

1. Introduction

The picophytoplankton (cyanobacteria and photosynthetic eukaryotes with cell diameter <3 μm) make up the smallest component of phytoplankton populations [1–3]. In the Arctic region, photosynthetic picoeukaryotes (PPEs) are major contributors to picophytoplankton and the small phytoplankton (<5 μm) represent 59%–63% of all marine photosynthetic biomass [4]. The Arctic has been undergoing accelerated warming and freshening since the 1990s due to the melting of multiyear sea ice and increasing river runoff into the Arctic Basin [5,6]. Environmental changes affect the phytoplankton and lead to an increase in PPEs contribution to total primary production and phytoplankton biomass [7–9].

Correct taxonomical identification of PPE requires the use of molecular methods. 18S rRNA gene-based environmental surveys have been increasingly used to investigate the composition of small eukaryotes. Molecular environmental studies conducted in Arctic and subarctic waters reported the presence of diverse microbial communities [10–17]. To better assess the diversity of small photosynthetic eukaryotes, flow cytometry via chlorophyll fluorescence was used to sort cells successfully [16,18–20]. However, the use of flow cytometry does not prevent the detection of heterotrophic eukaryotes that ingest photosynthetic organisms in their food vacuoles and thus could be detected by flow cytometry sorting that targeted chlorophyll fluorescence as well as the detection of photosynthetic algae with cell sizes larger than those of picoforms [20,21]. PPEs in under-ice waters remain understudied, especially for the season preceding the under-ice bloom in spring [10,11,17,22].

The White Sea is a small (area of 90,000 km^2 and volume of 6000 km^3) subarctic semi enclosed basin with an outlet to the Barents Sea. It has features similar to those of the Arctic shelf seas [23]. Usually from December to May the sea is covered with ice. The White Sea is strongly affected by continental runoff, and its waters are less saline (14–27 psu) than open ocean waters. The species composition and abundance of plankton algae have been studied for almost 80 years in the White Sea [24]. The species richness of nano- and microphytoplankton of the White Sea has been studied by microscopy and is represented by 450 taxa [24]. However, most of the studies examining the taxonomic diversity of the White Sea algae have been limited to the easily recognizable nano- and microsized algae, while the PPE composition remained understudied.

Previously, the taxonomic composition of PPEs was studied in summer plankton [25,26] and in the sea ice of the White Sea [27]. PPEs of the White Sea ice were represented by 16 algae genera belonging to eight classes and three supergroups. Chlorophyta, especially Mamiellophyceae, dominated among ice PPEs. The composition of the underlying ice waters' eukaryotic picophytoplankton in the White Sea was estimated for only one sample [28].

Considering the ongoing changes in the Arctic Ocean caused by global warming, and their implications, it is crucial to understand the PPE composition and provide detailed data on the prevalent taxa in subarctic waters. Hence, the objectives of this study were: i) to evaluate under-ice picophytoplankton abundance and biomass and the contribution of PPEs to total picophototroph abundance, ii) to reveal the taxonomic diversity of PPEs during the early spring by way of 18S rDNA sequencing, and iii) to compare the PPEs composition in under-ice water and ice. We targeted the study of the smallest size class of algae by a sample filtration approach, as they are abundant in the Arctic Ocean and difficult to identify by microscopy.

2. Materials and Methods

2.1. Sampling and Study Area

The samples were collected in Kandalaksha Bay, the White Sea on 19–23 March 2013 and 16–19 March 2014 near the White Sea Biological Station, Lomonosov Moscow State University (66°33′ N, 33°06′ E), from five different stations (Figure 1) with various degrees of under-ice water salinity and under-ice water current speeds (Table 1). The under-ice water salinity was lower due to the impact of the freshwater runoff at stations 1, 2, and 3 with the maximum freshening level at station 1. Under-ice water at stations 4 and 5 had salinity characteristics of the White Sea surface layer in the winter. The waters at stations 2, 4, and 5 are characterized by higher speeds of under-ice water currents than water areas of stations 1 and 3 [29]. Water samples at stations 1 and 2 were taken twice—in 2013 and 2014. The reference number of each sample consists of the station number, the last two digits of the year when the sample was taken, and the letter "w," which means water samples (e.g., 1/13w).

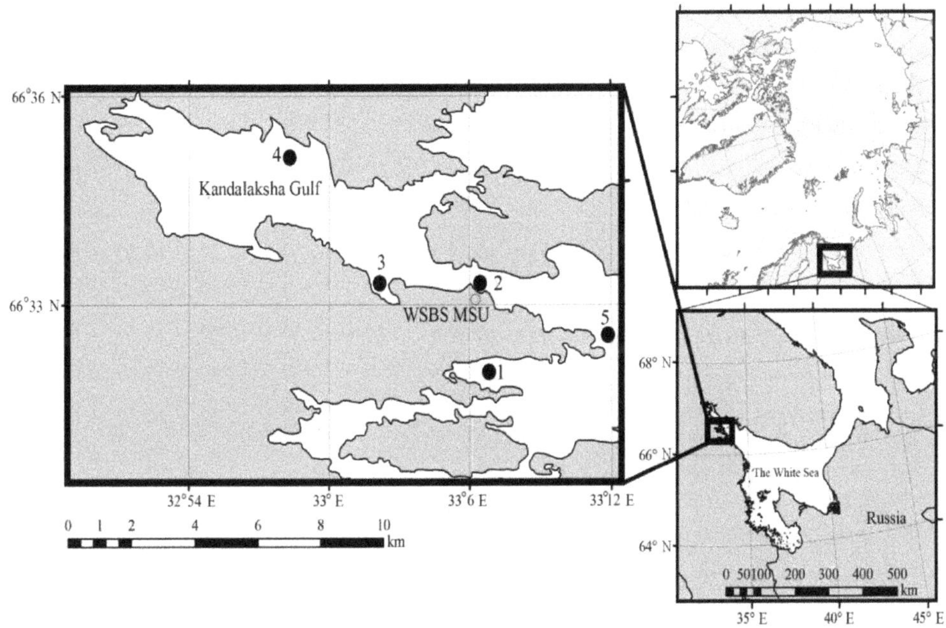

Figure 1. Location of the sampling stations in Kandalaksha Bay, White Sea in March 2013 and 2014.

Table 1. Characteristics of under-ice waters at the sampling stations of the White Sea: total chlorophyll *a* concentration, the abundance (N) and biomass (B) of photosynthetic picoeukaryotes, and their contribution to total picophytoplankton (including cyanobacteria) abundance (*N*, %) and biomass (B, %).

Sample	Latitude (N) Longitude (E)	Date	Ice Thickness (cm)	Under Ice Water		Chl-a (µg/L)	*N* (cells/mL)	B (µg C/L)	*N* (%)	B (%)
				Salinity (psu)	Temperature (°C)					
1/13w	66°32.01′	19 Mar 2013	71	14.9	−1.0	0.16	140	0.17	11	25
1/14w	33°6.54′	16 Mar 2014	58	15.6	−1.1	0.22	50	0.07	15	35
2/13w	66°33.20′	23 Mar 2013	22	24.5	−0.7	0.05	10	0.06	5	43
2/14w	33°6.28′	17 Mar 2014	49	24.5	−1.2	0.73	20	0.07	33	78
3/14w	66°33.12′ 33°2.11′	17 Mar 2014	52	21.9	−0.9	0.32	20	0.03	100	100
4/14w	66°34.87′ 32°58.89′	19 Mar 2014	45	26.7	−1.1	0.10	80	0.19	53	86
5/14w	66°32.14′ 33°13.17′	15 Mar 2014	26	25.5	−1.2	0.31	10	0.01	5	11

At each station, a titanium manual ice corer (14 cm of diameter) was used to make holes in the ice to collect 5 L of the underlying water. The ice thickness and water temperature were measured. Water temperature was measured directly with a probe Testo 108 (Testo, Lenzkirch, Germany). Within 1 h, water samples were brought to the laboratory, where the salinity was measured with a conductivity probe Cond 3150i (Xylem Analytics Germany Sales GmbH & Co. KG, WTW, Weilheim, Germany). The air temperature recordings from the weather station at the White Sea Biological Station were also used.

2.2. Chlorophylla

Subsamples of underlying water (500–1000 mL) were filtered through Whatman GF/F filters and frozen (−80 °C) for subsequent analysis. On returning to the Moscow laboratory, extractions and calculations were made following the procedure [30].

2.3. Enumeration of Picophototrophs

Whole (not prefiltered) seawater samples (10 mL) intended for analysis by epifluorescence microscopy were fixed with glutaraldehyde (AppliChem Panreac, Barcelona, Spain) at a final concentration of 1% (*v/v*). Nuclear filters (0.12 μm pore diameter) prestained with Sudan black were used for filtration. Cells with sizes <3 μm were enumerated at × 1000 magnification with a Leica DM2500 (Leica Microsystems GmbH, Wetzlar, Germany) epifluorescence microscope equipped with a 50 W mercury lamp under blue (Filter D; 355–425 nm) and green (Filter N2.1; 515–560 nm) excitation. The bright yellow fluorescence of the phycoerythrin-containing cyanobacteria could be distinguished easily from the deep red fluorescence of the chlorophyll-dominant picoeukaryotes. At least 300 cells at 30–50 microscopic fields were counted for each sample. Cell volumes were calculated as volumes of the relevant geometrical bodies [31] and then converted to their carbon content using the conversion factors of 470 fg C $cell^{-1}$ for prokaryotes and $0.433 \times (V)^{0.863}$pg C $cell^{-1}$ for PPEs [32].

2.4. DNA Isolation of Picoplanktonic Size-fraction

Three to five liters of water samples were filtered through a 2-μm pore size polycarbonate filter and then filtered again through 0.2-μm Sterivex units (Millipore Canada Ltd, Mississauga, ON, Canada). The buffer was added to the Sterivex units (1.8 mL of 50 mM Tris-HCl, 0.75 M sucrose, and 40 mM EDTA; pH 8.3). These units were stored at −80 °C until DNA extraction using the NucleoSpin Plant kit (Macherey-Nagel, Düren, Germany) following the manufacturer's instructions.

2.5. DNA Amplification and Sequencing

The ~0.43-kb fragments of the hypervariable V4 region of the 18S rRNA were amplified with the primer pair EuF-V4 (5′-CCAGCASCCGCGGTAATWCC-3′) and pico-R2 (5′-AKCCCCYAACTTTCGTTCTTGAT-3′) [27]. For PCR, the Encyclo Plus PCR kit (Evrogen, Moscow, Russia) was used. The volume of the amplification mixture was 30 μL. This was divided into three equal parts (10 μL each), and then PCR was carried out for each sample at three annealing temperatures, 55 °C, 60 °C, and 65 °C [27]. Cycling conditions were as follows: an initial denaturation step for 3 min at 94 °C, followed by 30 cycles (denaturation at 94 °C for 20 s, annealing for 20 s, and extension at 72 °C for 40 s), followed by a final extension at 72 °C for 5 min.

The PCR products obtained at three annealing temperatures were combined and, after extraction by agarose gel electrophoresis, were purified with a Cleanup Mini kit (Evrogen, Moscow, Russia). The resulting amplicons were used to prepare the libraries for the sequencing on the Illumina MiSeq platform (Illumina Inc., San Diego, CA, USA) with a TruSeq Nano DNA Kit (Illumina Inc., San Diego, CA, USA). The maximum read length of the Illumina MiSeq technology is about 500–600 bp, which matches and even exceeds the length of V4 of the SSU rRNA. The hypervariable V4 region of the 18S rRNA revealed an impressive hidden diversity in picoplanktonic communities [3,33].

The effective concentration of the libraries was tested by quantitative PCR with the primers I-qPCR-1.1 (5′-AATGATACGGCGACCACCGAGAT-3′) and I-qPCR-2.1 (5′-CAAGCAGAAGACGGCATACGA-3′). The library PhiX Control v3 (Illumina) was used as a control. Then the libraries were diluted to 12 pM and sequenced with a MiSeq Reagent Kit v.2 for 500 cycles. The pair-end read length was 250 × 2 bp.

2.6. Bioinformatics and Data Evaluation

The raw sequencing data were processed by Mothur software [34] and other procedures implemented in the SOP protocol [35]. Reads shorter than 150 bp and longer than 550 bp were removed, as well as reads with ambiguous bases (Ns) or >6 repeated bases. Assembled contigs were 430 bp in length, with ~70 bp overlapped paired reads. Identical sequences were removed by the unique.seqs command. The sequences were aligned using MAFFT with FFT-NS-2 strategy, a gap-opening penalty of 1.53, and a gap extension penalty of 0.123. Putative chimeric sequences were identified by UCHIME v 4.2.40 [36] and removed. A distance matrix of the high-quality sequences was constructed, and the sequences were clustered into operational taxonomic units (OTUs) at a 97% similarity level, with average neighbor clustering using cd-hit-est v,.3.1.2 [37]. The classification was performed by a local nucleotide BLAST search against the nonredundant version of the SILVA 123 SSU RNA database [38] using blastn (version 2.2.28+) with standard settings [39]. Sequences affiliated with nonprotist phyla or bacteria were eliminated. All singletons were removed. Consensus sequences of the OTUs were generated by the script described earlier [27]. All sequence reads were submitted to the GenBank BioProject (PRJNA368621) under the accession numbers MK571487-MK571523, MN541095, and MN684208. The phylogenetic tree was inferred by maximum likelihood method using RAxML 8.2.10 program [40], with default options according to GTRGAMMA model with 400 bootstrap replications, the number of which was set by bootstrapping criterion implemented in RAxML. The secondary structures of the terminal hairpins of V4 rRNA region were constructed according small subunit RNA secondary structure model [41].

Since the purpose of our research was limited to the photosynthetic picoeukaryotes, from the complete list of taxa revealed in under-ice water samples filtered through a 2-μm pore size filter, we chose only those species that have a cell size ≤3 μm. Where OTUs were identified to the genus/order/class level, we only analyzed taxa that, according to the published data [1,42,43], have species corresponding to the pico-size fraction. Since photosynthetic pico-sized cryptophytes were not detected microscopically and photosynthetic pico-sized dinoflagellates are not currently described, these groups were excluded from the analysis. Classes of algae are given according to AlgaeBase [44].

2.7. Statistical Procedures

The similarity matrix was calculated after standardization of the abundance of PPEs reads and square-root transformation for reducing the influence of the most dominant taxonomic entries [45]. The PRIMER v6 software (Primer-E Ltd, Plymouth, UK) [46] was used to group samples with similar taxonomic compositions by a group-average linkage cluster analysis and a nonmetric multidimensional scaling (MDS) ordination of a Bray–Curtis similarity matrix [45]. A breakdown of species similarities (SIMPER) was used to determine which taxon combination leads to the resulting groups [47].

3. Results

3.1. Environmental Conditions

The year 2014 was warmer than 2013 during the sampling period and the preceding month (Figure S1). The air temperature occasionally rose to the water freezing point or above, even in the middle of winter. Kandalaksha Bay was partially ice covered from December, with more complete ice cover in late March in both 2013 and 2014. Ice thickness varied from 22 cm to 71 cm (Table 1). Under-ice water salinity was lowest at station 1 in both years and varied between 14.9 psu and 15.6 psu (Table 1). At the other stations, the salinity of the under-ice water ranged from 21.9 psu to 26.7 psu. The temperature of the under-ice water varied very little: between −0.7 (2/13w) and −1.2 °C (2/14, 5/14w) with an average of −1.0 °C.

3.2. *Total Chlorophyll a Biomass*

Total chlorophyll *a* biomass (Chl *a*) level varied from 0.05 $\mu g \cdot L^{-1}$ at sample 2/13w to the highest value of 0.73 $\mu g \cdot L^{-1}$ at sample 2/14w (Table 1). The average Chl *a* concentration in under-ice water was 0.27 ± 0.21 $\mu g \cdot L^{-1}$.

3.3. *The Abundance of Picophototrophs*

The PPEs abundance ranged from 10 $cells \cdot mL^{-1}$ to 140 $cells \cdot mL^{-1}$ with an average of 50 $cells \cdot mL^{-1}$. The biomass varied between 0.03 and 0.17 $\mu g\ C \cdot L^{-1}$ (Table 1). Among photosynthetic pico-sized organisms, cyanobacteria dominated in all samples except 3/14w where we did not reveal photosynthetic prokaryotes. The relative abundance of PPEs varied significantly between 5% and 100% of the total cell counts and carbon biomass of pico-sized photosynthetic organisms.

3.4. *Taxonomic Composition of Eukaryotes in Samples Filtered through a 2-μm Pore Size Filter*

A total of 268,124 amplicons were sequenced from the seven samples, and 122,503 reads remained after quality filtering and preprocessing. The relative abundance of PPEs reads was 11%. The number of OTUs (at the 97% similarity level) that were clustered in individual samples varied between 609 and 3856 (Table 2).

Table 2. Summary of recovered reads and the number of operational taxonomic units (OTUs) in under-ice water picoplankton samples.

Sample	Total Number of V4 Tag Sequences	Number of V4 Sequences of Eukaryotes Groups after Quality Filtering	Number of Eukaryotes OTUs (97% Similarity)	Number of PPE Reads	Number of PPE OTUs
1/13w	26,493	7398	609	384	34
2/13w	24,600	19,080	1227	1542	98
1/14w	27,019	12,334	822	1347	72
2/14w	28,525	7645	618	1216	141
3/14w	39,578	17,845	1393	1620	83
4/14w	27,227	7974	738	1101	140
5/14w	94,682	50,227	3856	5947	186
Total	268,124	122,503	9263	13,157	754

Different OTUs are grouped according to their taxonomic affiliations to major phylogenetic groups, such as Chloroplastida, Stramenopiles (Bacillariophyta, Bolidophyceae, Chrysophyceae, Dictyochophyceae, Raphidophyceae, Pelagophyceae, Eustigmatophyceae), Alveolata, Rhizaria, Cryptophyta, Haptophyta, Opisthokonta, and others (Centrohelida, Telonemia, Kathablepharidae, Picozoa, and Eukaryota incertae sedis) (Figure 2). Protists from the taxonomic groups Rhizaria, Opisthokonta, Centrohelida, Telonemia, Kathablepharidae, and Picozoa are nonphotosynthetic forms. Alveolata and Cryptophyta include heterotrophic species. A total of 175 taxa of protists, determined to the genus level, and 148 forms, determined to higher taxonomic ranks, were found in water samples filtered through a 2-μm pore size filter.

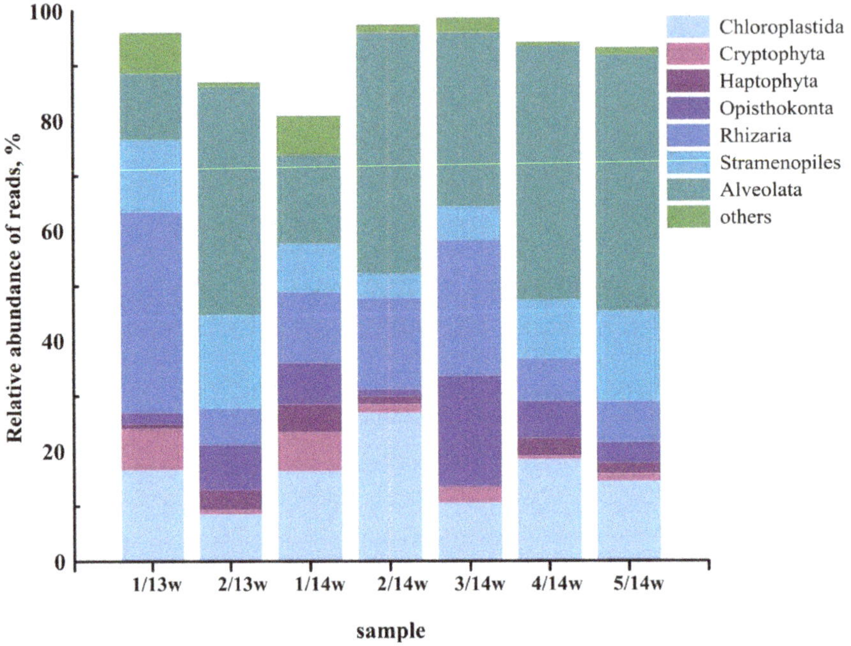

Figure 2. The relative abundance (%) of V4 rDNA reads of the major protist groups in picoplankton samples.

3.5. OTU Richness and Taxonomic Affiliation of the PPEs Sequences

PPEs belong to three supergroups: Chloroplastida (Chlorophyta), Stramenopiles, and Haptophyta (Table 3). Since different samples yielded different total numbers of sequence reads, they were normalized based on the lowest sample size (sample 1/13w—7398 reads) for comparing OTUs richness. The expected OTUs richness of PPEs was calculated with a 95% probability (Table 3). The minimum expected OTUs richness of PPEs species was observed at the lowest water salinity, in sample 1/13w. The highest expected richness was found at the highest values of salinity, in samples 4/14w and 5/14w.

Table 3. Relative abundance (%) of PPE groups based on V4 rDNA reads and the expected OTUs richness of PPEs per sample (95% probability). The standardized number of OTUs for each group is indicated in parentheses.

Taxonomic Group		Reads (%)						
		1/13w	2/13w	1/14w	2/14w	3/14w	4/14w	5/14w
Chloroplastida	Mamiellophyceae	81.2 (59)	27.3 (23)	71.3 (50)	91.8 (89)	82.4 (67)	60.8 (52)	70.2 (56)
	Pyramimonadophyceae	3.9 (6)	0.0	0.0	0.0	0.0	0.0	0.0
	Palmophyllophyceae	0.0	3.6 (3)	0.0	0.0	0.0	1.8 (2)	0.1 (1)
	Trebouxiophyceae	5.0 (6)	1.4 (4)	6.8 (8)	0.2 (1)	4.5 (7)	3.0 (6)	0.2 (1)
Stramenopile	Bolidophyceae	6.8 (21)	58.1 (67)	21.9 (42)	8.1 (12)	13.1 (24)	29.4 (36)	19.6 (39)
	Mediophyceae	0.0	0.0	0.0	0.0	0.0	0.0	0.2 (2)
Haptophyta	Coccolithophyceae	3.1 (9)	9.5 (3)	0.0	0.0	0.0	5.0 (4)	9.7 (2)
Expected OTUs richness of PPEs		36	71	57	105	63	110	103

3.6. Chlorophyta

Chlorophyta were represented by four classes: Mamiellophyceae, Pyramimonadophyceae, Palmophyllophyceae, and Trebouxiophyceae (Table 3). Mamiellophyceae were the predominant PPEs in all samples of under-ice water except sample 2/13w, where Bolidophyceae dominated. Within the Mamiellophyceae, *Micromonas* was the dominant genus in three samples and the *Mantoniella* genus dominated in four samples (Table 4). The negative correlation between the relative abundance of *Micromonas* and *Mantoniella* reads was found (Rs = −1; $p = 0.003$). The genera *Ostreococcus*, *Bathycoccus*, *Crustomastix* and OTU similar to the uncultured clone DSGM81 were also detected (Figure 3, Table 5).

Table 4. A relative abundance (%) of reads found for different taxonomical groups of Mamiellophyceae based on the V4 region of the 18S rRNA gene sequences.

Taxonomic Group	Reads (%)						
	1/13w	2/13w	1/14w	2/14w	3/14w	4/14w	5/14w
Micromonas polaris	1.6	17.1	1.7	54.9	5.4	66.7	53.1
Micromonas commoda A2	0.0	1.2	0.0	0.0	0.0	0.6	0.5
Micromonas clade F (B3)	0.0	0.7	0.0	0.0	0.0	0.0	0.4
total *Micromonas*	1.6	19.0	1.7	54.9	5.4	67.3	54.1
Mantoniella squamata	85.2	36.4	11.2	7.9	1.4	4.9	14.6
Mantoniella clade 1	7.7	4.2	42.4	25.8	5.4	8.7	7.1
Mantoniella clade 2	1.3	0.0	13.0	0.6	11.7	0.0	0.0
Mantoniella clade 3	3.5	24.6	30.8	1.4	75.9	8.7	4.5
Mantoniella clade 4	0.0	1.4	0.0	0.0	0.0	0.0	0.0
Total *Mantoniella*	97.7	66.6	97.4	35.8	94.4	22.3	26.2
Bathycoccus prasinos	0.0	4.7	0.0	0.3	0.0	3.3	2.1
Ostreococcus tauri	0.0	0.0	0.0	0.0	0.0	0.0	0.1
uncultured eukaryotic clone DSGM-81	0.6	9.2	0.9	8.1	0.2	6.7	14.0
Crustomastix sp. MBIC10709	0.0	0.5	0.0	1.0	0.0	0.4	3.6

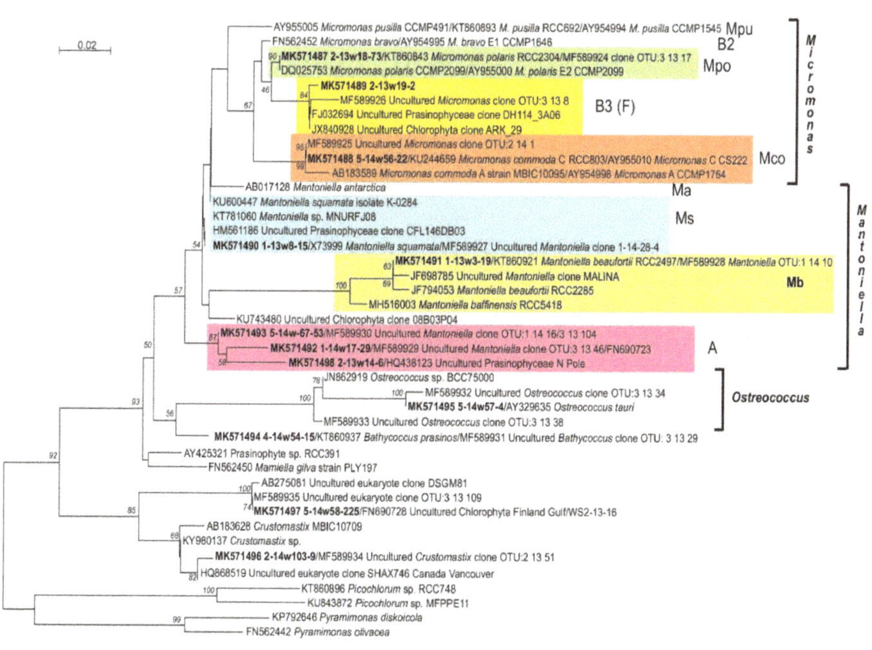

Figure 3. Phylogenetic tree of revealed Mamiellophyceae OTUs and Genbank reference sequences constructed from the V4 region of the 18S rRNA gene sequences by the maximum likelihood method. Bootstrap supporting values >0.5 are indicated. The scale is a number of nucleotide substitutions per site. Clades are designated according to Tragin and Vaulot (2019) [43] and Belevich et al. (2018) [27].

Table 5. The most abundant PPEs OTUs recovered in the under-ice water of Kandalaksha Bay, the White Sea in March 2013 and 2014 (clustering at 97% similarity threshold). The number of reads of each OTU is indicated in parentheses.

OTU	Closest Match	Identity	Origin
	Mamiellophyceae		
2/13w18-73 MK571487	*Micromonas pusilla* CCMP2099 (clade Ea) AY955000 (*Micromonas polaris*)	100%	Baffin Bay, Canada
5/14w56-22 MK571488	*Micromonas pusilla* strain CS222 (clade C) AY955010 (*Micromonas commoda* A2)	100%	South Pacific Ocean
2/13w19-2 MK571489	Uncultured *Micromonas* clade F MF589926	100%	the White Sea ice
	Uncultured Prasinophyceae clone DH114_3A06 FJ032694 (*Micromonas* clade B3)	100%	South Atlantic Ocean
1/13w8–15 MK571490	*Mantoniella squamata* X73999	100%	-
	Uncultured Prasinophyceae clone CFL146DB03 HM561186	100%	the Beaufort Sea
1/13w3–19 MK571491	*Mantoniella beaufortii* KT860921	100%	the Beaufort Sea
	Uncultured *Mantoniella* clade 1 MF589928	100%	the White Sea ice
1/14w17-29 MK571492	Uncultured *Mantoniella* clade 2 MF589929	100%	the White Sea ice
	Uncultured Chlorophyta clone 5-D5 FN690723	100%	the Baltic Sea
5/14w67-53 MK571493	Uncultured *Mantoniella* clade 3 MF589930	100%	the White Sea ice
	Uncultured Chlorophyta clone 5-D5 FN690723	98.4%	the Baltic Sea
2/13w14-6 MK571498	Uncultured Prasinophyceae clone North_Pole_SI120_29 HQ438123 (uncultured Mantoniella clade 4)	100%	North Pole sea ice
4/14w54–15 MK571494	*Bathycoccus prasinos* strain RCC801 KT860937	100%	English Channel, Atlantic Ocean
5/14w57-4 MK571495	*Ostreococcus tauri* Y15814	100%	the Mediterranean Sea
2/14w103-9 MK571496	Uncultured *Crustomastix* MF589934	100%	the White Sea ice
	Uncultured eukaryote clone SHAX746 HQ868519	99.4%	Pacific Ocean, Canada
5/14w58-225 MK571497	Uncultured eukaryotic clone DSGM81 AB275081	99.2%	methane cold seep sediment (Japan)
	Uncultured Chlorophyta FN690728	100%	the Baltic Sea
	Pyramimonadophyceae		
1/13w72–12 MK571500	*Pyramimonas* sp. RCC2009 JF794047	98.1%	the Beaufort Sea
	Palmophyllophyceae		
4/14w79–16 MK571499	*Prasinoderma coloniale* strain RCC854 KT860905	97.0%	Pacific Ocean
	Uncultured eukaryote clone SHAX501 HQ868998	98.9%	Pacific Ocean, Canada
	Trebouxiophyceae		
1/14w39-6 MK571501	*Picochlorum* sp. RCC748 KT860896	100%	Atlantic Ocean
1/14w40-46 MK571502	*Choricystis minor* X89012	100%	lake in Germany
	Bolidophyceae		
2/13w b284-3 MK571511	*Triparma strigata* KR998402	100%	Pacific Ocean, Japan
2/13w b270-21 MK571512	Uncultured bolidophyte LC190998	99.0%	Pacific Ocean, Japan
2/13w b264-63 MK571513	Uncultured bolidophyte LC191051	99.0%	Pacific Ocean, Japan
3/14w b337-71 MK571514	Uncultured stramenopile FN690655	99.0%	the Baltic Sea ice
1/14w b286-6 MN684208	Uncultured stramenopile FN690656	100%	the Baltic Sea ice
5/14w b537-30 MK571516	Uncultured eukaryote KT818381	97.8%	the Greenland Sea
4/14w b371-20 MK571518	Uncultured eukaryote KT811782	99.3%	the Greenland Sea
3/14w b346-24 MK571515	Uncultured eukaryote KT814386	98.8%	the Greenland Sea

Table 5. *Cont.*

OTU	Closest Match	Identity	Origin
5/14w b565-83 MK571517	Uncultured eukaryote KT815972	97.8%	the Greenland Sea
2/13w b265-11 MK571519	Uncultured eukaryote KT813573	99.5%	the Greenland Sea
4/14w b360-4 MK571520	Uncultured bolidophyte OTU:b474 MF589906	100%	The White Sea summer water
	Uncultured bolidophyte LC191049	100%	Pacific Ocean
2/13w b282-3 MK571522	Uncultured bolidophyte isolate OTU:b55 MF407369	100%	The White Sea ice
	Uncultured eukaryote clone 52c_105508 KT814907	99.5%	the Greenland Sea
4/14w b397-2 MK571523	Uncultured bolidophyte isolate OTU:b407 MF407373	100%	The White Sea summer water
	Mediophyceae		
5/14w120-8 MN541095	*Chaetoceros* cf. *neogracilis* strain RCC2318 JN934684	100%	the Beaufort Sea
5/14w103-6 MK571504	*Skeletonema marinoi* isolate 17 KR091067	100%	Atlantic Ocean
5/14w121-3 MK571505	*Minutocellus polymorphus* NIES-3970 LC189088	100%	-
	Arcocellulus cornucervis strain RCC2270 JN934677	100%	the Beaufort Sea
	Haptophyta		
5/14w130-550 MK571506	*Phaeocystis pouchetii* isolate AJ01 KR091066	100%	the North Sea
5/14w125-4 MK571507	*Chrysochromulina* clone MALINA JF698782	98.4%	the Beaufort Sea
	Uncultured eukaryote KP405041	99.2%	the South China Sea
2/13w29-6 MK571508	*Chrysochromulina simplex* AM491021	99.4%	-
	Uncultured haptophyte Ma135-Pry1-C55 JX680441	100%	the Marmara Sea
4/14w91-6 MK571509	Uncultured haptophyte FN690514	98.7%	the Baltic Sea
5/14w134-14 MK571510	Uncultured haptophyte KC488456	99.2%	the North Atlantic Ocean

The *Micromonas* genus was represented by two species and a recently described clade, corresponding to clade F [27] or clade B3 [43]. *Micromonas polaris* (previously *M. pusilla* clade Ea) was revealed in all samples; its contribution to the total *Micromonas* reads varied between 90% (2/13w) and 100% (1/13w, 1/14w, 2/14w, 3/14w). *Micromonas commoda* clade A2 [43] and *Micromonas* clade F (B3) were revealed much less often—in only three samples (2/13w, 4/14w and 5/14w) and two samples (2/13w and 5/14w), respectively. The contribution of each to the total *Micromonas* reads was low (Table 4).

There were several *Mantoniella* phylotypes from the three clades Ms, Mb, and A in the under-ice water of the White Sea (Figure 3). Three phylotypes were found in all samples: the first was identical to *Mantoniella squamata* (X73999), the second was matched to *Mantoniella beaufortii* (KT860921), and the third from clade A [43] was similar (>98%) to the Uncultured Chlorophyta clone 5-D5 (FN690723) from the Baltic Sea. In general, three phylotypes assigned to clade A were discovered in our samples. Two of them (MK571493 and MK571492) were previously found in the ice of the White Sea and identical to environmental sequences *Mantoniella* MF589930 and *Mantoniella* MF589928, respectively. The third *Mantoniella* phylotype (MK571498), with 100% similarity to the Uncultured Prasinophyceae clone North Pole SI120_29 (HQ438123) from the marine ice, was revealed in sample 2/13w. Substitutions in basal parts of helixes E23_1 and H 25 of 18S rRNA are diagnostic for distinguishing the phylotypes *M. squamata* and *M. beaufortii* and three other phylotypes of clade A, MK571493, MK571492, and MK571498 (Figure 4).

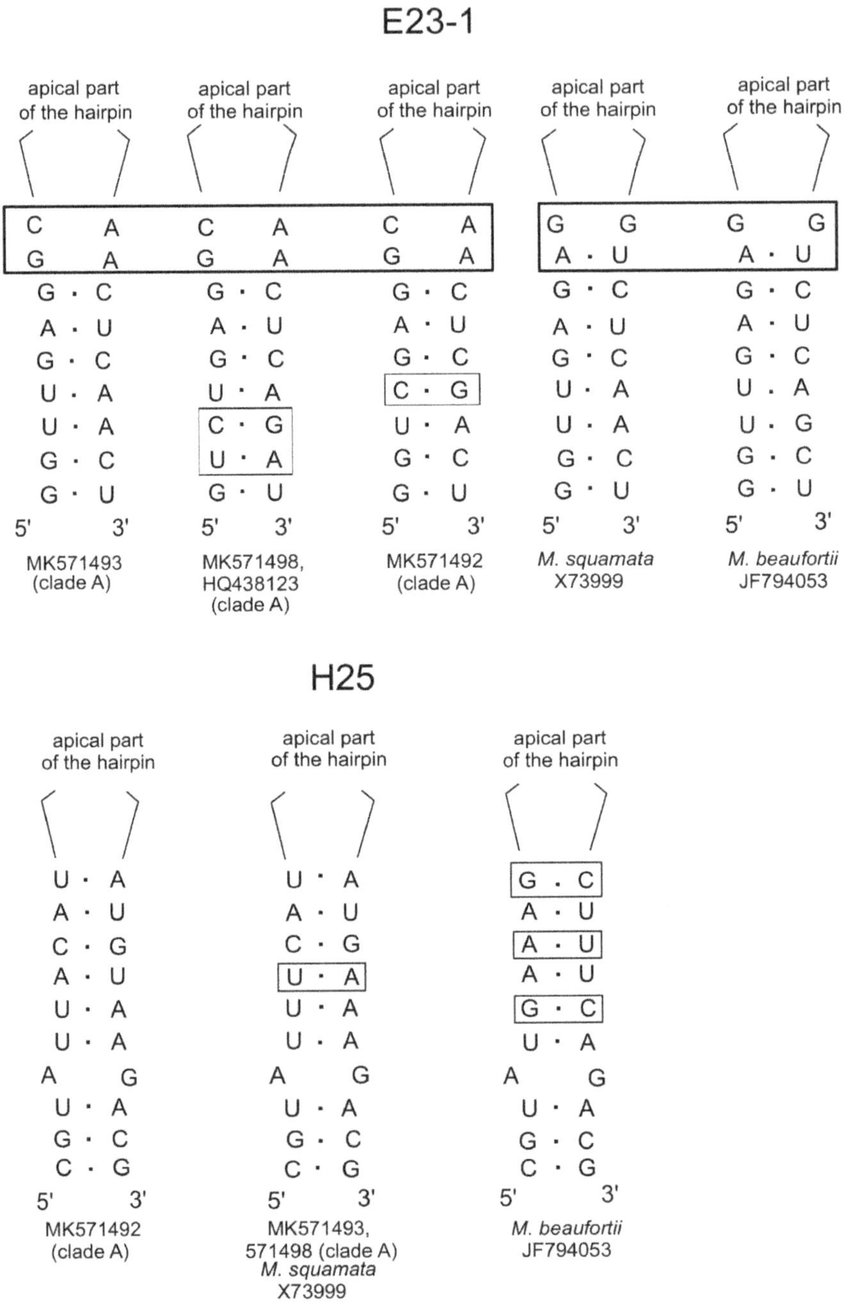

Figure 4. Compensatory base changes in the helices of the 18S rRNA secondary structure of *Mantoniella* (helices E23_1 and H25). The CBCs are shown in rectangles.

Bathycoccus OTUs found in samples 2/13w, 2/14w, 4/14w, and 5/14w were identical (100%) to the *Bathycoccus prasinos* strain RCC801 (KT860937). The relative read abundance of *B. prasinos* did not exceed 5% of Mamiellophyceae reads. The OTU matching *Ostreococcus tauri* (Y15814) was revealed only in sample 5/14w with a low (<1%) relative read abundance. Moreover, in all samples, we revealed OTUs that showed 99.2% similarity to the uncultured eukaryotic clone DSGM-81 (AB275081). The previous molecular phylogenetic analysis revealed that clone DSGM-81 belongs to Mamiellophyceae [27]. *Crustomastix* OTUs were found in four samples and showed 100% similarity to environmental sequences of uncultured *Crustomastix* (MF589934) previously identified in the White Sea ice.

Pyramimonadophyceae sequences were represented by Pyramimonas OTU, which is similar to Pyramimonas sp. (JF794047) from the Beaufort Sea. Among Palmophyllophyceae OTU, Prasinoderma similar to the *Prasinoderma coloniale* strain RCC854 was identified. Trebouxiophyceae from samples 1/13w, 1/14w, and 2/13w were identical to the freshwater algae *Choricystis minor* (X89012). Additionally, in all samples Trebouxiophyceae was represented by OTU identical to *Picochlorum* sp. RCC748 (KT860896) from the Atlantic Ocean.

3.7. Stramenopiles

Stramenopiles were represented by two classes: Bolidophyceae and Mediophyceae. All diatoms reads were revealed only in sample 5/14w with the highest sequencing depth. Among Mediophyceae, OTU identical to sequences of two different species, *Minutocellus polymorphus* (LC189088) and *Arcocellulus cornucervis* (JN934677), were revealed. *Skeletonema marinoi* (KR091067) and *Chaetoceros* cf. *neogracilis* (JN934684, KT860998) were also identified.

Bolidophyceae were revealed in all samples (Table 3, Figure 5). They were represented by sequences of *Triparma strigata* (KR998402) with 100% similarity, and OTUs similar to three phylotypes of uncultured bolidophytes earlier revealed in the ice and summer plankton of the White Sea [48], two phylotypes from the plankton of the Pacific Ocean (LC191051, LC190998), two uncultured stramenopiles from the Baltic Sea ice (FN690655, FN690656), and five uncultured eukaryotes from the Greenland Sea (KT818381, KT811782, KT814386, KT815972, and KT813573). Bolidophytes were the predominant PPEs in sample 2/13w.

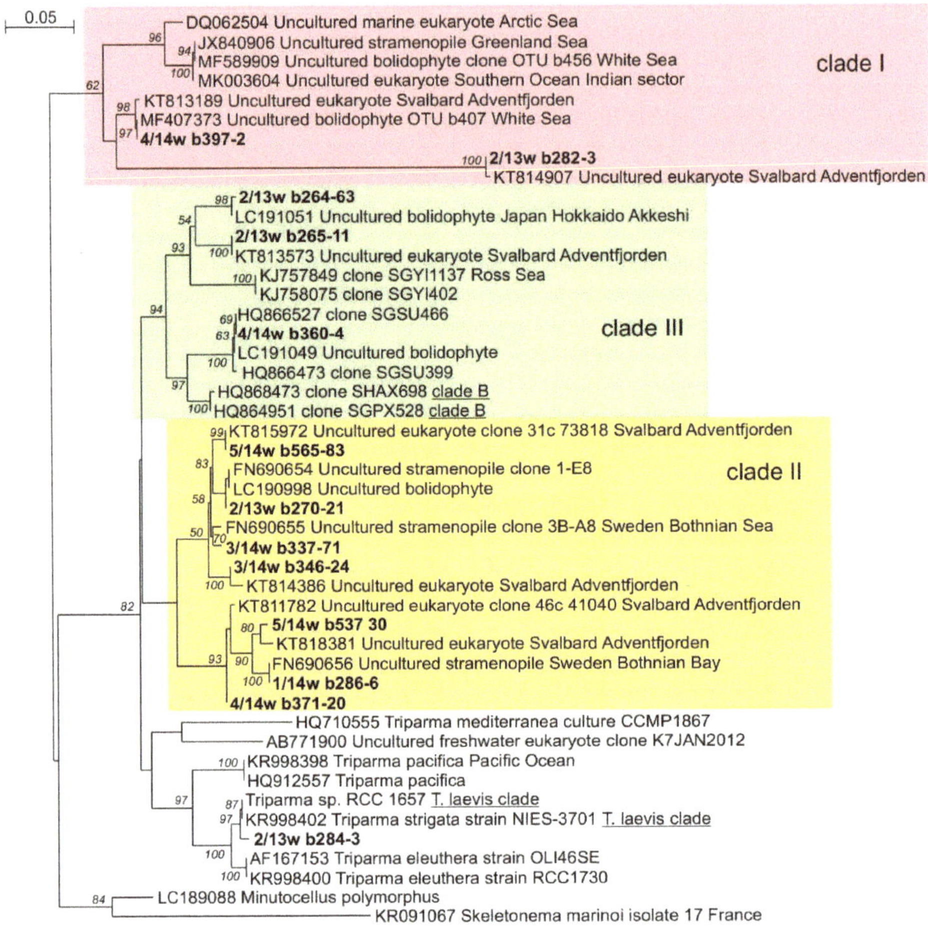

Figure 5. The maximum likelihood Bolidophyceae phylogenetic tree constructed from the V4 region of the 18S rDNA. Bootstrap support values >50% are indicated. The scale is the number of nucleotide substitutions per site.

3.8. *Haptophyta*

The contribution of Haptophyta reads varied between 3% and 10% (Table 3). Among the Haptophyta, two genera of class Coccolithophyceae were found—*Phaeocystis* and *Chrysochromulina*. *Phaeocystis* OTUs were similar (>99%) to the *Phaeocystis pouchetii* isolate AJ01 (KR091066) from the North Sea. Sequences closely related to the uncultured *Chrysochromulina* clone MALINA (JF698782) from the Beaufort Sea, occurred at insignificant levels only in the 5/14w sample. *Chrysochromulina simplex* (AM491021) was only found in sample 2/13w. Moreover, two Haptophyta phylotypes were classified at the phylum level with >98% similarity: the uncultured haptophyte (FN690514) from the Baltic Sea and the uncultured haptophyte (KC488456) from the North Atlantic Ocean.

3.9. *PPEs Community Structure*

Contributions of different taxa OTU reads to total numbers of PPE reads resulted in grouping of the stations into two clusters: CI at 64% similarity (1/13w, 1/14w and 3/14w) and CII at 62% similarity

(4/14w, 2/13w, 2/14w and 5/14w) (Figure 6). SIMPER analysis revealed that cluster C1 was characterized by a high contribution of *Mantoniella* reads to the total number of PPEs reads (53%), and cluster C2 was formed by stations with a high contribution of *Micromonas* (22%).

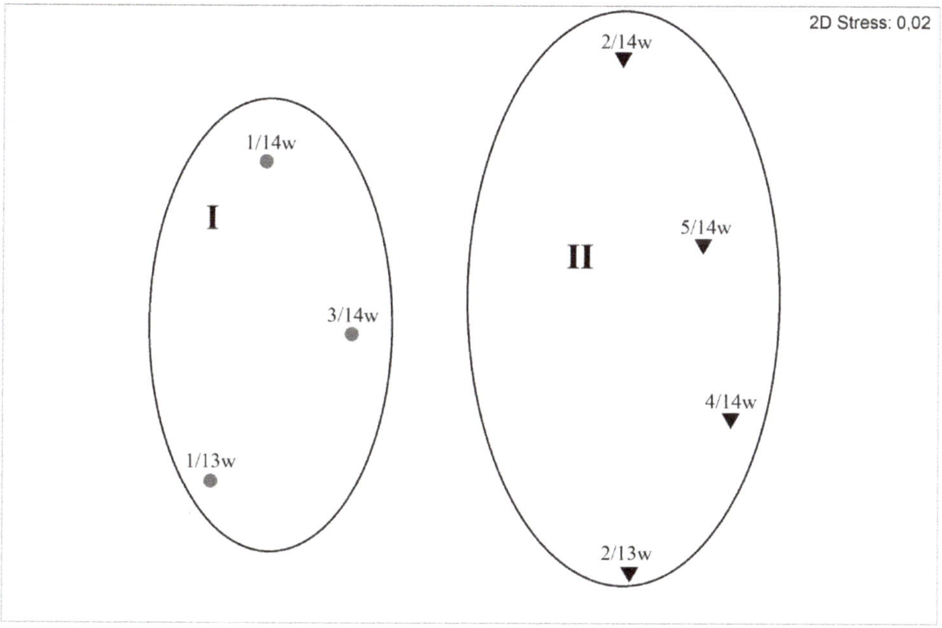

Figure 6. Community comparison of PPE assemblages at the different sampling stations using nonmetric multidimensional scaling (MDS) of a data matrix based on Bray-Curtis similarity.

4. Discussion

Our study revealed the most complete genetic diversity of under-ice PPEs in the White Sea, a unique marine environment combining features of temperate and Arctic seas. Such uniqueness of the abiotic environment was reflected in the composition of pico-sized photosynthetic organisms: besides widespread taxa (*Bathycoccus prasinos*, *Ostreococcus tauri*), the Arctic endemic *Micromonas polaris* (previously *M. pusilla* clade E2) and temperate waters *Micromonas commoda* A2 and *Mantoniella* were revealed. Temperate-water taxa survive in the White Sea despite extreme environmental conditions under the ice, i.e., near-freezing temperatures, polar night, and low irradiance, because of the snow and ice cover.

In the under-ice water of Kandalaksha Bay, Chl *a* concentrations in March 2013–2014 were double the values recorded in water samples taken directly underneath the ice of Chupa Inlet of Kandalaksha Bay in February 2002, but half the Chl *a* values in the same inlet in April 2002 [49]. This indicates that our studies of under-ice plankton algae were carried out in the prebloom period. Studies of the biomass plankton algae dynamics from January to April in Kandalaksha Bay also revealed the highest values in April [50]. The values of Chl *a* concentration obtained in the under-ice water of the White Sea are comparable to those in the under-ice water of the Baltic Sea in March (0.5–1.0 $\mu g\ L^{-1}$) [51].

The abundance of photosynthetic picoeukaryotes in our study was lower than the total number of PPEs observed in ice-covered underlying waters of Kandalaksha Bay (near station 2 in this study) in April 2010 [52]. At the same time, the average PPE abundance was comparable to the numbers of small photosynthetic eukaryotes (<2 μm) found in the under-ice water of Franklin Bay in December-March, but was approximately one order of magnitude lower than the abundance observed there in April-May [53].

In the under-ice water, PPEs were most abundant in three samples, while cyanobacteria dominated in the biomass in the other four samples. Earlier, the dominance of cyanobacteria was revealed in the summer plankton of the Onega [25] and Kandalaksha [52] bays of the White Sea.

Eukaryotic picoplankton is phylogenetically very diverse and includes lineages not yet described [1,54,55]. High-throughput sequencing of 18S rDNA of pico-sized eukaryotes from under-ice water yielded a detailed view of the plankton PPE community in the ice-covered underlying waters of the subarctic sea. As is commonly found in picoplankton diversity studies based on filtration approaches, sequences from larger protists and metazoans were recovered, probably due to cell breakage and the deformation of flexible walled cells, allowing their DNA and RNA to pass through the 3-μm filters [14,17,27,56]. However, some factors should be noted that can potentially lead to distortions in the estimation of real picoplankton diversity when using filtration. Among them are a probable breakage and deformation of larger cells [14,17,27,56], the presence of extracellular DNA in filtrates [57–59], and inaccurate size-based fraction separation. The metagenomic approach also has its limitations because of possible overestimation of particular OTUs due to high rRNA gene copy numbers or artifacts of the sequencing procedure [60,61], and insufficient 18S rDNA V4 region sequence resolution for detection of all morphospecies, as has been shown for diatoms, for example [62,63]. The exclusion of dinoflagellate sequences from the analysis due to the lack of cultured representatives of this group with cell sizes of less than 3 μm could affect the accuracy of our PPE diversity estimates [1]. Considering the above, 16 algae genera from seven classes and three supergroups are detected among the White Sea under-ice PPEs. Mamiellophyceae dominated in most of the samples. Palmophyllophyceae and Mediophyceae were the minor component of PPEs, Bolidophyceae made a significant contribution and even predominated in one sample.

Most sequences were assigned to Chlorophyta OTUs. Chlorophyta reads were abundant in mid-April during the early phase of the spring bloom in Norwegian waters (Isfjorden, West Spitsbergen) [64]. The dominance of Chlorophyta sequences was repeatedly noted in summer plankton communities of temperate and arctic waters [22,65–67]. On the class level, most Chlorophyta sequences were assigned to Mamiellophyceae OTUs, among which *Micromonas* and *Mantoniella* reads dominated. *Micromonas* was represented by species *Micromonas polaris*, *M. commoda* A2, and a phylotype of a recently described clade F according to Belevich et al. [27], or B3 according to Tragin and Vaulot [43]. *Micromonas polaris* dominated among *Micromonas*. *M. polaris* is widespread and dominant in the under-ice and open Arctic waters [11,17,43,68,69] and regarded as arctic endemic [11,19]. Previously, we detected *M. polaris* (as *Micromonas* E2) in the ice of the White Sea [24]. Its presence in the under-ice water and summer plankton of the subarctic White Sea and the Gulf of Finland of the subarctic Baltic Sea [43] once again indicates that the area of distribution of this species is wider than previously thought [11]. This endemic *M. polaris* does not seem to show intraspecific variability [70].

Micromonas commoda was detected earlier in the White Sea ice and summer plankton as *Micromonas* clade C [25,27]. This study is the first to show *M. commoda* in under-ice water. Its relative contribution to total *Micromonas* reads varied from 0% to 6% and was lower than in summer plankton [25]. *M. commoda* has ubiquitous distribution [43,71,72] and constitutes <1%–40% of Mamiellophyceae reads in different regions [43]. Unlike *M. polaris*, which does not show any intraspecific variability, the genetic diversity within *M. commoda* was previously highlighted [43,73,74] and it was suggested that speciation events might be ongoing within this species [72].

This study is the first to discover *Micromonas* clade F (B3) in the under-ice water of the White Sea. Previously, the phylogenetic analysis of Mamiellophyceae revealed the existence of a new clade, *Micromonas* F, in the ice and summer plankton of the White Sea [25,27]. Later, the analysis of the taxonomic diversity and global distribution of *Micromonas* revealed the existence of *Micromonas* clade B3 [43]. This subarctic clade combined amplicons that are 100% similar to OTUs of *Micromonas* clade F from the White Sea ice. *Micromonas* clade B3 made a great contribution to Mamiellophyceae reads from Canada (32%) and amounted to more than 10% of Mamiellophyceae reads at four subarctic stations off Maine and Iceland, as well as at a temperate location off the U.K. coast in the North Sea [43].

The contribution of this taxon to the total Mamiellophyceae reads in under-ice plankton was <1%, which is much lower than in summer plankton [25].

The results of this work represent the first sighting of a broad diversity of *Mantoniella* phylotypes in the under-ice water (Figure 3): *Mantoniella* OTUs match *M. squamata* (clade Ms), *M. beaufortii* (clade Mb), and three other *Mantoniella* lineages from clade A [43] found earlier in the White Sea ice [27]. It has been suggested that clade A is potentially an ice alga [43]. This assumption does not agree with the fact that *Mantoniella* phylotype MK571493 dominated among the Mamiellophyceae reads in two out of seven samples of under-ice water and did not dominate in any of the ice samples [27]. The *Mantoniella* MK571493 and MK571492 are 98.4% similar; however, out of six substitutions making those two phylotypes different from each other, four are compensatory. Likewise, the *Mantoniella* MK571498 sequence is 97.3%–97.8% similar to two other sequences from the same A clade. Substitutions in basal parts of helixes E23_1 and H 25 are differentiated in all known *Mantoniella* phylotypes (Figure 4). Further research may lead to the discovery of new species in the *Mantoniella* A clade.

OTUs of *Ostreococcus* found in the White Sea plankton were identical to the sequence of *O. tauri* Y15814, which was isolated from the Thau lagoon (Mediterranean Sea) with highly variable salinity from 24 to 38 psu [75]. It was suggested that *O. tauri* might represent several species, adapted to different degrees of salinity [43]. The relative abundance of *O. tauri* in under-ice water (<0.1%) was significantly lower than that in summer plankton (31%) [25].

The relative abundance of *Bathycoccus prasinos* reads in under-ice water of the White Sea did not exceed 5%, whereas in summer plankton it reached 33% [25]. *B. prasinos* is a widespread alga with global distribution from tropical to polar waters [43,76]. *Bathycoccus* is now known to be composed of two cryptic species with identical 18S rRNA sequences but differences in the ITS, as well as at the genomic level [76,77]. One of them could be coastal, while another might have adapted to warmer oceanic waters [76–78].

In under-ice water, Palmophyllophyceae and Pyramimonadophyceae were represented only by one taxon each; their respective contributions to total PPEs reads were low. The cell size of *Pyramimonas* sp. (JF794047) is unknown and its assignment to picoforms may be inappropriate. In our samples, several *Pyramimonas* taxa with nanosizes were found (*Pyramimonas mucifera*, *P. olivacea*, *P. tetrarhynchus*, etc.), which is consistent with the high diversity of nanoforms of this genus in arctic and subarctic plankton [19,79].

Trebouxiophyceae were represented by both freshwater *Choricystis minor* and marine *Picochlorum* sp. algae. Its contribution to the total PPEs reads was low (Table 3) and exceed 5% only at station 1 (samples 1/13w and 1/14w), which was most affected by river flows (Table 1). Earlier, the dominance of Trebouxiophyceae reads was revealed in the ice at this station [27].

Diatoms made an insignificant contribution to the PPE community and were found only in sample 5/14 with the maximum sequencing depth. Among all identified diatom taxa, we can confidently assert that only *Minutocellus polymorphus/Arcocellulus cornucervis* and *Skeletonema marinoi* match the picofraction. Unfortunately, the resolving power of V4 is insufficient to correctly identify *M. polymorphus* and *A. cornucervis*—their V4 regions are identical [63]. Earlier, *M. polymorphus* was not recorded in the under-ice plankton communities; like *S. marinoi*, it was registered in the sympagic communities of the White Sea [27]. The revealed OTU of *Chaetoceros* cf. *neogracilis* matches two culture representatives of this species deposited in the RCC culture collection (Roscoff, France) numbers RCC2318 and RCC2507. The cell size of both algae is more than 3 μ. However, small *Chaetoceros* are abundant in spring bloom waters, and the simple morphology could hide a high diversity of species [80,81].

Our study identified a limited variety of Haptophyta in the White Sea under-ice water. We found *Phaeocystis pouchetii* and *Chrysochromulina simplex* that are widespread in the plankton of the subarctic and Arctic seas [19,51,82,83]. The cell sizes of the identified *Chrysochromulina* sp. and two uncultured haptophytes are unknown, but they supposedly can match the size of picofraction. Egge et al. [83] discovered six OTUs assigned to *Chrysochromulina* that were only found in the picoplankton size fraction rather than the nanoplankton. Since Haptophyta DNA is known to amplify poorly, the molecular methods can underestimate these algae diversity [16,84]. In the Arctic, the number of haptophyte OTUs found in plankton fraction <3 μm varies significantly, between 10–12 in Beaufort Sea and English Channel [16,19] and 59 in the North Pacific [85].

The phylogenetic analysis of Bolidophyceae sequences from the under-ice plankton of the White Sea showed the highest diversity of these algae among the identified PPE classes. Bolidophytes are considered true picosized forms [86]. Bolidophytes were represented by *Triparma strigata* and 12 OTUs of uncultured forms. A large number of bolidophyte sequences are uncultured forms, as noted earlier for the Arctic or subarctic locations, as well as for the Baltic Sea [14,65,79,82,87]. Previously, ice only two phylotypes of bolidophytes were found in the White Sea [27]. Based on a phylogenetic analysis, the number of White Sea bolidophytes was increased by including sequences from the Genbank that had not been previously classified as Bolidophyceae and had been deposited at the NCBI Genbank as uncultured stramenopile or uncultured eukaryotes [47]. In the under-ice water of the White Sea, four OTUs of bolidophytes have 100% similarity with OTUs found earlier in the ice (1 OTU) and summer plankton (2 OTUs). The phylogenetic analysis recovered that the White Sea bolidophytes refer to three environmental clades (env. clades I, II and III [88] (Figure 4)), in addition to the group corresponding to the genus *Triparma*. Bolidophyte *T. strigata* has a worldwide distribution in plankton and is most abundant in polar waters [89–91]. The complete sequences of the 18S rRNA gene are almost identical (similarity of 99.9%–100%) for such morphologically distinct species as *T. strigata*, *T. laevis f. longispina*, *T.* aff. *verrucosa*, and flagellate *Triparma* sp. RCC1657; therefore, if only the 18S RNA gene sequences are considered, all those species are combined in the clade of *T. laevis* [90]. Therefore, the discovery of a phylotype similar to *T. strigata* does not necessarily mean that there is only this species in the under-ice plankton of the White Sea. Bolidophyceae sequences dominated among PPEs reads in sample 2/13. This is the first registration of Bolidophyceae domination in the subarctic plankton.

The dominance of *Micromonas polaris* reads in three out of seven samples corresponds to the fact that our studies were carried out in the prebloom period. *M. polaris* had exceptionally high relative read abundances during pre- and postbloom stages in Isfjorden, West Spitsbergen [64], the Amundsen Gulf, and the Canadian Beaufort Sea [68]. An unexpected result was the high share of different *Mantoniella* taxa in three samples and Bolidophyceae in one sample. Situations where the relative read abundance of *M. polaris* was lower than other taxa were noted earlier in different regions of the Arctic and subarctic: the unexpectedly high proportion of *Bathycoccus* was revealed in July surface samples in the Amundsen Gulf and Canadian Beaufort Sea [68]. It was suggested that this might have been associated with offshore upwelling, or, more speculatively, a viral attack on *Micromonas* triggered by specific oceanographic conditions. *Mantoniella squamata* made a great contribution to the Mamiellophyceae reads off Greenland [43]. The dominance of *Mantoniella* from clade A in PPE reads was identified for the first time. The spatial variability of relative read abundances may be controlled by the combined influence of abiotic and biotic factors.

Earlier, at the same stations as in the present work, we studied the taxonomic composition of PPE and protists in ice samples filtered through a 2-μm pore size filter [27]. Different assemblages colonized the under-ice water and the ice. In samples of under-ice water filtered through a 2-μm pore size filter, Stramenopiles made the most significant contribution to total quality reads (average: 34%), whereas Rhizaria dominated in the ice samples (average: 18%). At the same time, the number of identified protists taxa determined to genus level was comparable: 175 in water and 185 in ice. In the Baltic Sea, the ice community was more diverse than the wintertime water [79]. The contributions of Chloroplastida in ice and water were comparable; the Alveolata contribution was lower in ice than in under-ice water. It is interesting to note that Alveolates were not the dominant group in any of the samples of the White Sea plankton, while the domination of Alveolates in the reads abundance was noted in size-fractionated seawater (0.2–3.0 μm) of the Amundsen Gulf flaw lead system [13].

The similarity of PPE composition in under-ice water and ice was 0.75 (Sørensen index). The variety of phylotypes of certain genera in plankton was lower than in ice. For example, the genus *Ostreococcus* in plankton was represented only by *O. tauri*, while *O. tauri* and *Ostreococcus* sp. were recorded in ice. Pelagophyceae was found in ice but not in under-ice water. Chlorophyta dominated in both habitats, but the contributions of specific genera and classes varied. For example, the domination of *Mantoniella* in water (three stations) was not observed in the ice, where *Micromonas* always made the most significant contribution. In addition, the domination of Bolidophyceae was observed only in under-ice water (one sample), while the domination of Trebouxiophyceae was only found in ice (one station).

The research undertaken on White Sea under-ice photosynthetic picoeukaryotes' genetic diversity is one stage in studying the dynamics of plankton communities in the subarctic.

Supplementary Materials: The following are available online at http://www.mdpi.com/1424-2818/12/3/93/s1. Figure S1. Air temperature at the White Sea Biological Station (shaded area corresponds to sampling period).

Author Contributions: Conceptualization, T.A.B. and L.V.I.; methodology, M.D.L. and I.A.M.; validation, T.A.B., L.V.I. and A.V.T.; formal analysis, T.A.B., M.D.L. and A.V.T.; investigation, I.A.M. and T.A.B.; resources, T.A.B. and A.V.T.; data curation, I.A.M. and T.A.B.; writing—original draft preparation, T.A.B.; writing—review and editing, A.V.T. and L.V.I.; visualization, A.V.T. and T.A.B.; supervision, A.V.T.; project administration, A.V.T.; funding acquisition, T.A.B. and A.V.T. All authors have read and agreed to the published version of the manuscript.

Funding: This study was done in the framework of the Moscow University Project "Noah's Ark" and of the State Tasks of the Lomonosov Moscow State University (themes no. AAAA-A16-116021660052-0, and AAAA-A17-117120540067-0), with the financial support of the Russian Foundation for Basic Research (project No. 16-05-00502).

Conflicts of Interest: The authors declare that they have no conflict of interest.

References

1. Vaulot, D.; Eikrem, W.; Viprey, M.; Moreau, H. The diversity of small eukaryotic phytoplankton (<3 μm) in marine ecosystems. *FEMS Microbiol. Rev.* **2008**, *32*, 795–820. [CrossRef] [PubMed]
2. Massana, R. Eukaryotic Picoplankton in Surface Oceans. *Annu. Rev. Microbiol.* **2011**, *65*, 91–110. [CrossRef] [PubMed]
3. Moon-van der Staay, S.Y.; De Wachter, R.; Vaulot, D. Oceanic 18S rDNA sequences from picoplankton reveal unsuspected eukaryotic diversity. *Nature* **2001**, *409*, 607–610. [CrossRef] [PubMed]
4. Bhavya, P.A.S.; Lee, J.H.; Lee, H.W.; Kang, J.J.; Lee, J.H.; Lee, D.; An, S.H.; Stockwell, D.A.; Whitledge, T.E.; Lee, S.H. First in situ estimations of small phytoplankton carbon and nitrogen uptake rates in the Kara, Laptev, and East Siberian seas. *Biogeosciences* **2018**, *15*, 5503–5517. [CrossRef]
5. Proshutinsky, A.; Krishfield, R.; Timmermans, M.-L.; Toole, J.; Carmack, E.; McLaughlin, F.; Williams, W.J.; Zimmermann, S.; Itoh, M.; Shimada, K. Beaufort Gyre freshwater reservoir: State and variability from observations. *J. Geophys. Res.* **2009**, *114*, 1–25. [CrossRef]
6. Carmack, E.; Mclaughlin, F. Towards recognition of physical and geochemical change in Subarctic and Arctic Seas. *Prog. Oceanogr.* **2011**, *90*, 90–104. [CrossRef]
7. Li, W.K.W.; McLaughlin, F.A.; Lovejoy, C.; Carmack, E.C. Smallest algae thrive as the Arctic Ocean freshens. *Science* **2009**, *326*, 539. [CrossRef]
8. Vincent, W.F. Microbial ecosystem responses to rapid climate change in the Arctic. *ISME J.* **2010**, *9*, 1087–1090. [CrossRef]
9. Comeau, A.M.; Li, W.K.W.; Tremblay, J.É.; Carmack, E.C.; Lovejoy, C. Arctic ocean microbial community structure before and after the 2007 record sea ice minimum. *PLoS ONE* **2011**, *6*, e27492. [CrossRef]
10. Lovejoy, C.; Massana, R.; Pedros-Alio, C. Diversity and distribution of marine microbial eukaryotes in the Arctic Ocean and adjacent seas. *Appl. Environ. Microbiol.* **2006**, *72*, 3085–3095. [CrossRef]
11. Lovejoy, C.; Vincent, W.F.; Bonilla, S.; Roy, S.; Martineau, M.-J.; Terrado, R.; Potvin, M.; Massana, R.; Pedros-Alio, C. Distribution, phylogeny, and growth of cold-adapted picoprasinophytes in arctic seas. *J. Phycol.* **2007**, *43*, 78–89. [CrossRef]
12. Hamilton, A.K.; Lovejoy, C.; Galand, P.E.; Ingram, R.G. Water masses and biogeography of picoeukaryote assemblages in a cold hydrographically complex system. *Limnol. Oceanogr.* **2008**, *53*, 922–935. [CrossRef]
13. Terrado, R.; Vincent, W.F.; Lovejoy, C. Mesopelagic protists: Diversity and succession in a coastal Arctic ecosystem. *Aquat. Microb. Ecol.* **2009**, *56*, 25–39. [CrossRef]

14. Terrado, R.; Medrinal, E.; Dasilva, C.; Thaler, M.; Vincent, W.; Lovejoy, C. Protist community composition during spring in an Arctic flaw lead polynya. *Polar Biol.* **2011**, *34*, 1901–1914. [CrossRef]
15. Lovejoy, C.; Potvin, M. Microbial eukaryotic distribution in a dynamic Beaufort Sea and the Arctic Ocean. *J. Plankton Res.* **2011**, *33*, 431–444. [CrossRef]
16. Marie, D.; Shi, X.L.; Rigaut-Jalabert, F.; Vaulot, D. Use of flow cytometric sorting to better assess the diversity of small photosynthetic eukaryotes in the English Channel. *FEMS Microbiol. Ecol.* **2010**, *72*, 165–178. [CrossRef]
17. Meshram, A.R.; Vader, A.; Kristiansen, S.; Gabrielsen, T.M. Microbial eukaryotes in an Arctic under-ice spring bloom north of Svalbard. *Front. Microbiol.* **2017**, *8*, 1099. [CrossRef] [PubMed]
18. Kirkham, A.R.; Jardillier, L.E.; Tiganescu, A.; Pearman, J.; Zubkov, M.V.; Scanlan, D.J. Basin-scale distribution patterns of photosynthetic picoeukaryotes along an Atlantic Meridional Transect. *Environ. Microbiol.* **2011**, *13*, 975–990. [CrossRef]
19. Balzano, S.; Marie, D.; Gourvil, P.; Vaulot, D. Composition of the summer photosynthetic pico and nanoplankton communities in the Beaufort Sea assessed by T-RFLP and sequences of the 18S rRNA gene from flow cytometry sorted samples. *ISME J.* **2012**, *6*, 1480–1498. [CrossRef]
20. Kataoka, T.; Yamaguchi, H.; Sato, M.; Watanabe, T.; Taniuchi, Y.; Kuwata, A.; Kawachi, M. Seasonal and geographical distribution of near-surface small photosynthetic eukaryotes in the western North Pacific determined by pyrosequencing of 18S rDNA. *FEMS Microbiol. Ecol.* **2017**, *93*, fiw229. [CrossRef]
21. Diez, B.; Pedros-Alio, C.; Massana, R. Study of genetic diversity of eukaryotic picoplankton in different oceanic regions by small-subunit rRNA gene cloning and sequencing. *Appl. Environ. Microbiol.* **2001**, *67*, 2932–2941. [CrossRef] [PubMed]
22. Metfies, K.; von Appen, W.-J.; Kilias, E.; Nicolaus, A.; Nöthig, E.-M. Biogeography and photosynthetic biomass of Arctic marine picoeukaroytes during summer of the record sea ice minimum 2012. *PLoS ONE* **2016**, *11*, e0148512. [CrossRef] [PubMed]
23. Berger, V.; Dahle, S.; Galaktionov, K.; Kosobokova, X.; Naumov, A.; Rat'kova, T.; Savinov, V.; Savinova, T. *White Sea: Ecology and Environment*; Derzavets Publisher: St-Petersburg-Tromsø, Russia, 2001.
24. Ilyash, L.V.; Belevich, T.A.; Zhitina, L.S.; Radchenko, I.G.; Ratkova, T.N. Phytoplankton of the White Sea. In *Biogeochemistry of the Atmosphere, Ice and Water of the White Sea: The White Sea Environment Part I*; Lisitzin, A., Gordeev, V., Eds.; Springer: Cham, Switzerland, 2018. [CrossRef]
25. Belevich, T.A.; Ilyash, L.V.; Milyutina, I.A.; Logacheva, M.D.; Troitsky, A.V. Phototrophic picoeukaryotes of Onega Bay, the White Sea: Abundance and species composition. *Mosc. Univ. Biol. Sci. Bull.* **2017**, *72*, 109–114. [CrossRef]
26. Milyutina, I.A.; Belevich, T.A.; Ilyash, L.V.; Troitsky, A.V. Insight into picophytoplankton diversity of the subarctic White Sea—The first recording of Pedinophyceae in environmental DNA. *MicrobiologyOpen* **2019**, *8*, e892. [CrossRef] [PubMed]
27. Belevich, T.A.; Ilyash, L.V.; Milyutina, I.A.; Logacheva, M.D.; Goryunov, D.V.; Troitsky, A.V. Photosynthetic picoeukaryotes in the land-fast ice of the White Sea, Russia. *Microb. Ecol.* **2018**, *75*, 582–597. [CrossRef] [PubMed]
28. Belevich, T.A.; Ilyash, L.V.; Milyutina, I.A.; Logacheva, M.D.; Goryunov, D.V.; Troitsky, A.V. Metagenomic analyses of White Sea picoalgae: First data. *Biochemistry* **2015**, *80*, 1514–1521. [CrossRef]
29. Pantyulin, A.N. Dynamics, structure, and water masses. In *The White Sea System: Water Column and Interacting Atmosphere, Cryosphere, the River Run-Off, and Biosphere*; Lisitzin, A.P., Ed.; Scientific World: Moscow, Russia, 2012; Volume 2, pp. 309–379. (In Russian)
30. Arar, E.J.; Collins, G.B. Method 445.0. In *In Vitro Determination of Chlorophyll A and Pheophytin A in Marine and Freshwater Algae by Fluorescence*; U.S. Environmental Protection Agency: Washington, DC, USA, 1997.
31. Hillebrand, H.; Dürselen, C.D.; Kirschtel, D.; Pollingher, U.; Zohary, T. Biovolume calculation for pelagic and benthic microalgae. *J. Phycol.* **1999**, *5*, 403–424. [CrossRef]
32. Verity, P.G.; Robertson, C.Y.; Tronzo, C.R.; Endrews, M.G.; Nelson, J.R.; Sieracki, M.E. Relationship between cell volume and the carbon and nitrogen content of marine photosynthetic nanoplankton. *Limnol. Oceanogr.* **1992**, *37*, 1434–1446. [CrossRef]
33. López-García, P.; Rodríguez-Valera, F.; Pedrós-Alió, C.; Moreira, D. Unexpected diversity of small eukaryotes in deep-sea Antarctic plankton. *Nature* **2001**, *409*, 603–607. [CrossRef]

34. Schloss, P.D.; Westcott, S.L.; Ryabin, T.; Hall, J.R.; Hartmann, M.; Hollister, E.B.; Lesniewski, R.A.; Oakley, B.B.; Parks, D.H.; Robinson, C.J.; et al. Introducing mothur: Open-source, platform-independent, community-supported software for describing and comparing microbial communities. *Appl. Environ. Microbiol.* **2009**, *75*, 7537–7541. [CrossRef]
35. Kozich, J.J.; Westcott, S.L.; Baxter, N.T.; Highlander, S.K.; Schloss, P.D. Development of a dual-index sequencing strategy and curation pipeline for analyzing amplicon sequence data on the MiSeq Illumina sequencing platform. *Appl. Environ. Microbiol.* **2013**, *79*, 5112–5120. [CrossRef]
36. Edgar, R.C.; Haas, B.J.; Clemente, J.C.; Quince, C.; Knight, R. UCHIME improves sensitivity and speed of chimera detection. *Bioinformatics* **2011**, *27*, 2194–2200. [CrossRef] [PubMed]
37. Li, W.; Godzik, A. Cd-hit: A fast program for clustering and comparing large sets of protein or nucleotide sequences. *Bioinformatics* **2006**, *22*, 1658–1659. [CrossRef] [PubMed]
38. Quast, C.; Pruesse, E.; Yilmaz, P.; Gerken, J.; Schweer, T.; Yarza, P.; Peplies, J.; Glöckner, F.O. The SILVA ribosomal RNA gene database project: Improved data processing and web-based tools. *Nuc. Acids Res.* **2013**, *41*, D590–D596. [CrossRef] [PubMed]
39. Camacho, C.; Coulouris, G.; Avagyan, V.; Ma, N.; Papadopoulos, J.; Bealer, K.; Madden, T. BLAST+: Architecture and applications. *BMC Bioinform.* **2009**, *10*, 421. [CrossRef] [PubMed]
40. Stamatakis, A. RAxML version 8: A tool for phylogenetic analysis and post-analysis of large phylogenies. *Bioinformatics* **2014**, *30*, 1312–1313. [CrossRef] [PubMed]
41. Wuyts, J.; De Rijk, P.; De Peer, V.Y.; Pison, G.; Rousseeuw, P.; De Wachter, R. Comparative analysis of more than 3000 sequences reveals the existence of two pseudoknots in area V4 of eukaryotic small subunit ribosomal RNA. *Nucleic Acids Res.* **2000**, *28*, 4698–4708. [CrossRef]
42. Santos, A.L.; Pollina, T.; Gourvil, P.; Corre, E.; Marie, D.; Garrido, J.L.; Rodríguez, F.; Noël, M.-H.; Vaulot, D.; Eikrem, W. Chloropicophyceae, a new class of picophytoplanktonic prasinophytes. *Sci. Rep.* **2017**, *7*, 14019. [CrossRef]
43. Tragin, M.; Vaulot, D. Novel diversity within marine Mamiellophyceae (Chlorophyta) unveiled by metabarcoding. *Sci. Rep.* **2019**, *9*, 5190. [CrossRef]
44. Guiry, M.D.; Guiry, G.M. AlgaeBase. World-Wide Electronic Publication, National University of Ireland, Galway. 2017. Available online: http://www.Algaebase.Org (accessed on 1 May 2019).
45. Clarke, K.R.; Warwick, R.M. *Change in Marine Communities: An Approach to Statistical Analysis and Interpretation*; Primer-E Ltd.: Plymouth, UK, 2001.
46. Clarke, K.R.; Gorley, R.N. *PRIMER v6: User Manual/Tutorial*; Primer-E Ltd.: Plymouth, UK, 2006.
47. Clarke, K.R. Non-parametric multivariate analyses of changes in community structure. *Aust. J. Ecol.* **1993**, *18*, 117–143. [CrossRef]
48. Belevich, T.A.; Ilyash, L.V.; Milyutina, I.A.; Logacheva, M.D.; Troitsky, A.V. Metagenomics of Bolidophyceae in plankton and ice of the White Sea. *Biochemistry* **2017**, *82*, 1917–1928. [CrossRef] [PubMed]
49. Krell, A.; Ummenhofer, C.; Kattner, G.; Naumov, A.; Evans, D.; Dieckmann, G.S.; Thomas, D.N. The biology and chemistry of land fast ice in the White Sea, Russia – a comparison of winter and spring conditions. *Polar Biol.* **2003**, *26*, 707–719. [CrossRef]
50. Ilyash, L.V.; Zhitina, L.S.; Kudryavtseva, V.A.; Mel'nikov, I.A. Seasonal dynamics of algae species composition and biomass in the coastal ice of Kandalaksha Bay, the White Sea. *Biol. Bull. Rev.* **2012**, *73*, 459–470. (In Russian)
51. Majaneva, M.; Blomster, J.; Müller, S.; Autio, R.; Majaneva, S.; Hyytiäinen, K.; Nagai, S.; Rintala, J.-M. Sea-ice eukaryotes of the Gulf of Finland, Baltic Sea, and evidence for herbivory on weakly shade-adapted ice algae. *Eur. J. Protistol.* **2017**, *57*, 1–15. [CrossRef]
52. Belevich, T.A.; Ilyash, L.V. Picophytoplankton abundance in the Velikaya Salma strait, White Sea. *Microbiology* **2012**, *81*, 389–395. [CrossRef]
53. Terrado, R.; Lovejoy, C.; Massana, R.; Vincent, W.F. Microbial food web responses to light and nutrients beneath the coastal Arctic Ocean sea ice during the winter–spring transition. *J. Mar. Syst.* **2008**, *74*, 964–977. [CrossRef]
54. Epstein, S.; Lopez-Garcia, P. "Missing" protists: A molecular prospective. *Biodivers. Conserv.* **2008**, *17*, 261–276. [CrossRef]
55. Massana, R.; Pedros-Alio, C. Unveiling new microbial eukaryotes in the surface ocean. *Curr. Opin. Microbiol.* **2008**, *11*, 213–218. [CrossRef]

56. Sørensen, N.; Daugbjerg, N.; Gabrielsen, T.M. Molecular diversity and temporal variation of picoeukaryotes in two Arctic fjords, Svalbard. *Polar Biol.* **2012**, *35*, 519–533. [CrossRef]
57. Nielsen, K.M.; Johnsen, P.J.; Bensasson, D.; Daffonchio, D. Release and persistence of extracellular DNA in the environment. *Environ. Biosaf. Res.* **2007**, *6*, 37–53. [CrossRef]
58. Charvet, S.; Vincent, W.F.; Comeau, A.; Lovejoy, C. Pyrosequencing analysis of the protist communities in a high Arctic meromictic lake: DNA preservation and change. *Front. Microbiol.* **2012**, *3*, 422. [CrossRef] [PubMed]
59. Not, F.; del Campo, J.; Balagué, V.; de Vargas, C.; Massana, R. New Insights into the Diversity of Marine Picoeukaryotes. *PLoS ONE* **2009**, *4*, e7143. [CrossRef] [PubMed]
60. Zhu, F.; Massana, R.; Not, F.; Marie, D.; Vaulot, D. Mapping of picoeukaryotes in marine ecosystems with quantitative PCR of the 18S rRNA gene. *FEMS Microb. Ecol.* **2005**, *52*, 79–92. [CrossRef] [PubMed]
61. Potvin, M.; Lovejoy, C. PCR-based diversity estimates of artificial and environmental 18S rRNA gene libraries. *J. Eukaryot. Microbiol.* **2009**, *56*, 174–181. [CrossRef] [PubMed]
62. Luddington, I.; Kaczmarska, I.; Lovejoy, C. Distance and character-based evaluation of the V4 region of the 18S rRNA gene for the identification of diatoms (Bacillariophyceae). *PLoS ONE* **2012**, *7*, e45664. [CrossRef] [PubMed]
63. Balzano, S.; Percopo, I.; Siano, R.; Gourvil, P.; Chanoine, M.; Dominique, M.; Vaulot, D.; Sarno, D. Morphological and genetic diversity of Beaufort Sea diatoms with high contributions from the *Chaetoceros neogracilis* species complex. *J. Phycol.* **2017**, *53*, 161–187. [CrossRef]
64. Marquardt, M.; Vader, A.; Stübner, E.I.; Reigstad, M.; Gabrielsen, T.M. Strong seasonality of marine microbial eukaryotes in a high-Arctic fjord (Isfjorden, in West Spitsbergen, Norway). *Appl. Environ. Microbiol.* **2016**, *82*, 1868–1880. [CrossRef]
65. Kilias, E.S.; Nöthig, E.M.; Wolf, C.; Metfies, K. Picoeukaryote plankton composition off West Spitsbergen at the Entrance to the Arctic Ocean. *J. Eukaryot. Microbiol.* **2014**, *61*, 569–579. [CrossRef]
66. Monier, A.; Comte, J.; Babin, M.; Forest, A.; Matsuoka, A.; Lovejoy, C. Oceanographic structure drives the assembly processes of microbial eukaryotic communities. *ISME J.* **2015**, *9*, 990–1002. [CrossRef]
67. Zhang, F.; He, J.; Lin, L.; Jin, H. Dominance of picophytoplankton in the newly open surface water of the central Arctic Ocean. *Polar Biol.* **2015**, *38*, 1081–1089. [CrossRef]
68. Joli, N.; Monier, A.; Logares, R.; Lovejoy, C. Seasonal patterns in Arctic prasinophytes and inferred ecology of *Bathycoccus* unveiled in an Arctic winter metagenome. *ISME J.* **2017**, *11*, 1372–1385. [CrossRef]
69. Kilias, E.; Wolf, C.; Nöthig, E.M.; Peeken, I.; Metfies, K. Protist distribution in the western Fram Strait in summer 2010 based on 454-pyroseqeuncing of 18S rDNA. *J. Phycol.* **2013**, *49*, 996–1010. [CrossRef] [PubMed]
70. Balzano, S.; Gourvil, P.; Siano, R.; Chanoine, M.; Marie, D.; Lessard, S.; Sarno, D.; Vaulot, D. Diversity of cultured photosynthetic flagellates in the northeast Pacific and Arctic Oceans in summer. *Biogeosciences* **2012**, *9*, 4553–4571. [CrossRef]
71. Foulon, E.; Not, F.; Jalabert, F.; Cariou, T.; Massana, R.; Simon, N. Ecological niche partitioning in the picoplanktonic green alga *Micromonas pusilla*: Evidence from environmental surveys using phylogenetic probes. *Environ. Microbiol.* **2008**, *10*, 2433–2443. [CrossRef]
72. Simon, N.; Foulon, E.; Grulois, D.; Six, C.; Desdevises, Y.; Latimier, M.; Le Gall, F.; Tragin, M.; Houdan, A.; Derelle, E.; et al. Revision of the genus *Micromonas* Manton et Parke (Chlorophyta, Mamiellophyceae), of the type species *M. pusilla* (Butcher) Manton & Parke and of the species *M. commoda* van Baren, Bachy and Worden and description of two new species based on the genetic and phenotypic characterization of cultured isolates. *Protist* **2017**, *168*, 612–635. [CrossRef] [PubMed]
73. Slapeta, J.; López-García, P.; Moreira, D. Global dispersal and ancient cryptic species in the smallest marine eukaryotes. *Mol. Biol. Evol.* **2006**, *23*, 23–29. [CrossRef]
74. Worden, A.Z.; Lee, J.-H.; Mock, T.; Rouzé, P.; Simmons, M.P.; Aerts, A.L.; Allen, A.E.; Cuvelier, M.L.; Derelle, E.; Everett, M.V.; et al. Green evolution and dynamic adaptations revealed by genomes of the marine picoeukaryotes *Micromonas*. *Science* **2009**, *324*, 268–272. [CrossRef]
75. Courties, C.; Perasso, R.; Chrétiennot-Dinet, M.-J.; Gouy, M.; Guillou, L.; Troussellier, M. Phylogenetic analysis and genome size of *Ostreococcus tauri* (Chlorophyta, Prasinophyceae). *J. Phycol.* **1998**, *34*, 844–849. [CrossRef]

76. Vannier, T.; Leconte, J.; Seeleuthner, Y.; Mondy, S.; Pelletier, E.; Aury, J.-M.; de Vargas, C.; Sieracki, M.; Iudicone, D.; Vaulot, D.; et al. Survey of the green picoalga *Bathycoccus* genomes in the global ocean. *Sci. Rep.* **2016**, *6*, 37900. [CrossRef]
77. Vaulot, D.; Lepére, C.; Toulza, E.; De la Iglesia, R.; Poulain, J.; Gaboyer, F.; Moreau, H.; Vandepoele, K.; Ulloa, O.; Gavory, F.; et al. Metagenomes of the picoalga *Bathycoccus* from the Chile Coastal Upwelling. *PLoS ONE* **2012**, *7*, e39648. [CrossRef]
78. Limardo, A.J.; Sudek, S.; Choi, C.J.; Poirier, C.; Rii, Y.M.; Blum, M.; Roth, R.; Goodenough, U.; Church, M.J.; Worden, A.Z. Quantitative biogeography of picoprasinophytes establishes ecotype distributions and significant contributions to marine phytoplankton. *Environ. Microbiol.* **2017**, *19*, 3219–3234. [CrossRef] [PubMed]
79. Majaneva, M.; Rintala, J.M.; Piisila, M.; Fewer, D.P.; Blomster, J. Comparison of wintertime eukaryotic community from sea ice and open water in the Baltic Sea, based on sequencing of the 18S rRNA gene. *Polar Biol.* **2012**, *35*, 875–889. [CrossRef]
80. Booth, B.C.; Larouche, P.; Belanger, S.; Klein, B.; Amiel, D.; Mei, Z.P. Dynamics of *Chaetoceros socialis* blooms in the North Water. *Deep Sea Res. II Top. Stud. Oceanogr.* **2002**, *49*, 5003–5025. [CrossRef]
81. Degerlund, M.; Eilertsen, H.C. Main species characteristics of phytoplankton spring blooms in NE Atlantic and Arctic waters (68–80° N). *Estuar. Coast. Shelf Sci.* **2010**, *33*, 242–269. [CrossRef]
82. Dasilva, C.R.; Li, W.K.W.; Lovejoy, C. Phylogenetic diversity of eukaryotic marine microbial plankton on the Scotian Shelf, Northwestern Atlantic Ocean. *J. Plankton Res.* **2013**, *36*, 344–363. [CrossRef]
83. Egge, E.S.; Johannessen, T.V.; Andersen, T.; Eikrem, W.; Bittner, L.; Larsen, A.; Sandaa, R.A.; Edvardsen, B. Seasonal diversity and dynamics of haptophytes in the Skagerrak, Norway, explored by high-throughput sequencing. *Mol. Ecol.* **2015**, *24*, 3026–3042. [CrossRef]
84. Liu, H.; Probert, I.; Uitz, J.; Claustre, H.; Aris-Brosou, S.; Frada, M.; Not, F.; Vargas, C. Extreme diversity in noncalcifying haptophytes explains a major pigment paradox in open oceans. *Proc. Natl. Acad. Sci. USA* **2009**, *106*, 12803–12808. [CrossRef]
85. Orsi, W.; Song, Y.C.; Hallam, S.; Edgcomb, V. Effect of oxygen minimum zone formation on communities of marine protist. *ISME J.* **2012**, *6*, 1586–1601. [CrossRef]
86. Ichinomiya, M.; Yoshikawa, S.; Kamiya, M.; Ohki, K.; Takaichi, S.; Kuwata, A. Isolation and characterization of Parmales (Heterokonta/Heterokontophyta/Stramenopiles) from the Oyashio region, Western North Pacific. *J. Phycol.* **2011**, *47*, 144–151. [CrossRef]
87. Ichinomiya, M.; dos Santos, A.; Gourvil, P.; Yoshikawa, S.; Kamiya, M.; Ohki, K.; Audic, S.; de Vargas, C.; Noël, M.-H.; Vaulot, D.; et al. Diversity and oceanic distribution of the Parmales (Bolidophyceae), a picoplanktonic group closely related to diatoms. *ISME J.* **2016**, *10*, 2419–2434. [CrossRef]
88. Kuwata, A.; Yamada, K.; Ichinomiya, M.; Yoshikawa, S.; Tragin, M.; Vaulot, D.; dos Santos, A. Bolidophyceae, a Sister Picoplanktonic Group of Diatoms—A Review. *Front. Mar. Sci.* **2018**, *5*, 370. [CrossRef]
89. Komuro, C.; Narita, H.; Imai, K.; Nojiri, Y.; Jordan, R.W. Microplankton assemblages at Station KNOT in the subarctic western Pacific, 1999–2000. *Deep Sea Res. II Top. Stud. Oceanogr.* **2005**, *52*, 2206–2217. [CrossRef]
90. Ichinomiya, M.; Nakamachi, M.; Shimizu, Y.; Kuwata, A. Growth characteristics and vertical distribution of *Triparma laevis* (Parmales) during summer in the Oyashio region, western North Pacific. *Aquat. Microb. Ecol.* **2013**, *68*, 107–116. [CrossRef]
91. Piwosz, K.; Wiktor, J.M.; Niemi, A.; Tatarek, A.; Michel, C. Mesoscale distribution and functional diversity of picoeukaryotes in the first-year sea ice of the Canadian Arctic. *ISME J.* **2013**, *7*, 1461–1471. [CrossRef]

© 2020 by the authors. Licensee MDPI, Basel, Switzerland. This article is an open access article distributed under the terms and conditions of the Creative Commons Attribution (CC BY) license (http://creativecommons.org/licenses/by/4.0/).

diversity

MDPI

Article

Patterns in Alpha and Beta Phytoplankton Diversity along a Conductivity Gradient in Coastal Mediterranean Lagoons

Natassa Stefanidou [1,*], Matina Katsiapi [1,2], Dimitris Tsianis [3], Maria Demertzioglou [4], Evangelia Michaloudi [4] and Maria Moustaka-Gouni [1,*]

1 Department of Botany, School of Biology, Aristotle University of Thessaloniki, 54124 Thessaloniki, Greece
2 EYATH SA, Water Supply Division/Drinking Water Treatment Facility, Nea Ionia, 57008 Thessaloniki, Greece; matinakatsiapi@gmail.com
3 Fisheries Department, Dir. of Agricultural Economy & Veterinary Med., Regional District of Xanthi, Region of Eastern Macedonia & Thrace, 67100 Xanthi, Greece; tsianis75@yahoo.gr
4 Department of Zoology, School of Biology, Aristotle University of Thessaloniki, 54124 Thessaloniki, Greece; dememari92@gmail.com (M.D.); tholi@bio.auth.gr (E.M.)
* Correspondence: astefanid@bio.auth.gr (N.S.); mmustaka@bio.auth.gr (M.M.-G.)

Received: 16 December 2019; Accepted: 17 January 2020; Published: 19 January 2020

Abstract: Understanding the diversity patterns of phytoplankton assemblages in coastal lagoons is clearly important for water management. In this study, we explored alpha and beta diversity patterns in phytoplankton communities across five Mediterranean lagoons hydrologically connected to Vistonikos Gulf. We examined the phytoplankton community composition and biomass on a monthly basis from November 2018 to October 2019. For this, water samples were collected from seven inshore, brackish and coastal waters, sampling sites covering a wide range of conductivity. We found significant spatial and temporal differences in phytoplankton alpha diversity and in phytoplankton biomass metrics explained by the high variation of conductivity. Evenness remained low throughout the study period, reflecting significant dominance of several phytoplankton blooms. Harmful algal blooms of *Prorocentrum minimum*, *Alexandrium* sp., *Rhizosolenia setigera* and *Cylindrotheca closterium* occurred. The system's species pool was characterized by relatively high phytoplankton beta diversity (average ~0.7) resulting from high temporal species turnover (90%). Overall, alpha and beta diversity components were indicative of rather heterogeneous phytoplankton communities which were associated with the high differences in conductivity among the sampling sites.

Keywords: brackish; HABs; dinoflagellates; diatoms; cyanobacteria; transitional water bodies

1. Introduction

Coastal lagoons are shallow semi-enclosed dynamic ecosystems which are connected to the sea by one or more restricted inlets [1]. They are transitional zones between the terrestrial and the marine environment occupying 13% of the coastal areas worldwide [2]. Coastal lagoons provide essential ecosystem services [3]. Due to their shallow depth and the restricted water exchange with the adjacent sea, they are naturally high productive ecosystems [4,5] that are usually subjected to intensive human activities, mostly aquaculture, with several consequences [6–8]. In addition to human exploitation, coastal lagoons have been identified as one of the most vulnerable habitants to potential impacts associated with climate change [5,9]. Projections for atmospheric temperature increase up to 6 °C in parallel with the more often and stronger hot extremes for the Mediterranean region [10,11] would most probably increase the salinity in coastal lagoons affecting, in turn, the biological communities. Studies using microbial communities as experimental systems have pointed out that a potential salinity change in coastal Mediterranean environment would significantly affect the microbial biodiversity, biomass and resource use efficiency [12,13].

Within the Mediterranean basin, there are numerous (>100) coastal lagoons of varying size [14,15]. Multiple line of evidence suggest that these ecosystems host an important fraction of global biodiversity including rare species [16–18]. A considerable number of studies examining the alpha diversity patterns (i.e., the local diversity within a site) of macrophytes (e.g., [19]), fish (e.g., [20]) and benthic invertebrates (e.g., [21]) in coastal lagoons relate them to environmental parameters (e.g., total phosphorus). More recently, studies of the spatial alpha diversity gradients of phytoplankton communities relate them to geographical variables (e.g., altitude in [22]) or other physical (e.g., water temperature in [23]) and chemical (e.g., nutrients in [24]) environmental aspects. Furthermore, investigations were carried out in order to determine the key environmental factors driving the frequent phytoplankton blooms in these shallow productive ecosystems [25]. The advancement of Next Generation Sequencing (NGS) tools in recent years has allowed the investigation of deep microbial diversity in coastal lagoons, revealing information for their alpha diversity patterns in relation to other biotic or abiotic parameters (e.g., [26–28]).

However, still today, very little is known about the microbial (including phytoplankton) beta diversity (i.e., the ratio of the local diversity divided by the regional diversity) and how it varies along the conductivity gradient in such ecosystems. Whittaker et al. [29] defined beta diversity as the extent of differentiation among biological communities along habitat gradients within an ecosystem. In particular, beta diversity is a measure of the difference in species composition between two or more local assemblages (e.g., the five Mediterranean lagoons under study) [30]. There are two potential ways in which two or more assemblages can be different: one is species replacement (i.e., turnover) and the second is species loss or gain, which implies that the poorest assemblage is a subset of the richest one (nestedness) [31]. The differentiation of the spatial turnover and nestedness components of beta diversity is crucial for understanding central biogeographic, ecological and conservation issues [32].

In the present study we explored the patterns of phytoplankton biodiversity across five Mediterranean lagoons which are located nearby and are hydrologically connected to Vistonikos Gulf, North Aegean Sea. To the best of our knowledge, this is the first study on these lagoons' phytoplankton. Particularly, we examined the spatial and temporal differences in phytoplankton species richness and other alpha diversity estimators (i.e., Shannon index) along the conductivity gradient that characterizes these lagoons. Furthermore, we explored the degree of homogeneity across the different waterbodies by computing beta diversity. By calculating nestedness and turnover, we examined whether beta diversity originated from differences in richness level (nestedness) or from species replacement (turnover). Finally, in order to study the effect of conductivity on phytoplankton communities we modelled their relationship using different phytoplankton attributes (e.g., phytoplankton biomass, species and group composition richness). Overall, we aimed to evaluate whether hydrological connectivity through water intrusion from Vistonikos Gulf or environmental heterogeneity in terms of conductivity stronger affects the phytoplankton community structure across the five lagoons. In the first case we expect that phytoplankton communities would be relatively homogeneous (i.e., low beta diversity and low turnover) with no significant relationships between conductivity and the different phytoplankton attributes. Alternatively, in the case where conductivity would significantly affect the phytoplankton communities across the lagoons, we expect rather heterogeneous phytoplankton communities (i.e., high beta diversity) resulting from species turnover.

2. Materials and Methods

2.1. Sampling Sites and Sample Collection

Phytoplankton samples were collected monthly on an annual basis (November 2018 to October 2019) from five Mediterranean lagoons located in Thrace, Northeastern Greece. All the lagoons are connected to Vistonikos Gulf (North Aegean Sea) by canals (Table 1; Figure 1). In total, 84 phytoplankton samples were analyzed from seven inshore sites of these lagoons. Vistonis lagoon has a surface area of approximately 45 km^2 with a mean depth of 3 m and it is connected to Vistonikos Gulf through three tidal canals [33]. Being the largest lagoon among the five studied, samples were collected from three different sites in the southern part of Vistonis lagoon nearby Vistonikos Gulf. The second largest lagoon, Lagos lagoon, has a surface area of 3 km^2 and mean depth of 2 m. In its northern part it communicates with the southern part of Vistonis lagoon while in its southern part with Vistonikos Gulf through tidal canals [34]. Lafra and Lafrouda are smaller lagoons than Vistonis and Lagos and have approximately 1.2 km^2 and 0.8 km^2 surface areas, respectively, and a mean depth <0.5 m [35]. The smallest lagoon is Palaia Koiti lagoon which occupies a surface area of <0.1 km^2 and has a mean depth of approximately 1 m [35]. Lafra, Lafrouda and Palaia Koiti lagoons are connected with Vistonikos Gulf through tidal canals.

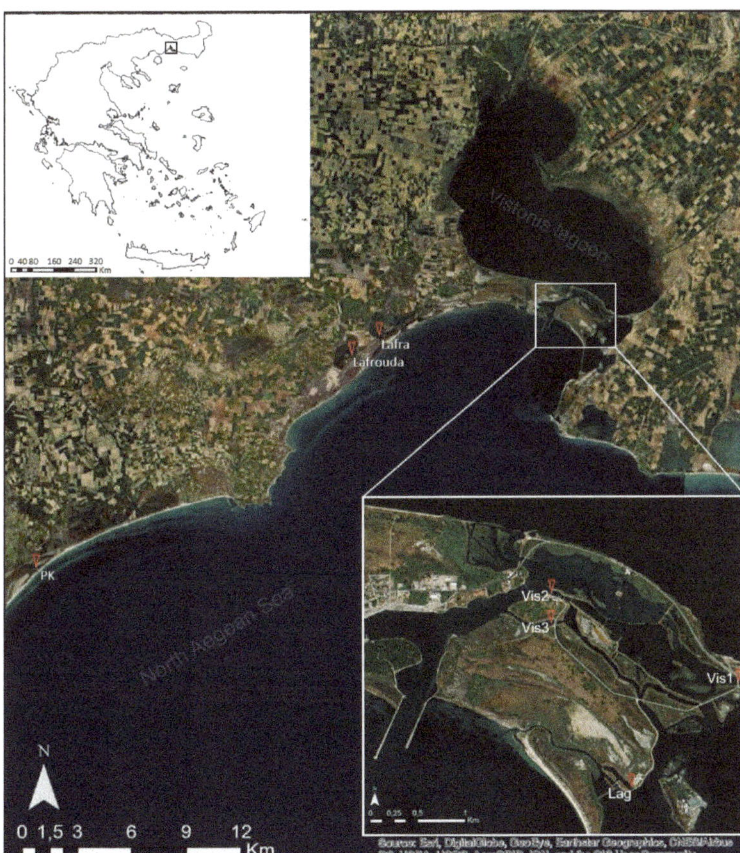

Figure 1. Study area and sampling sites in Lafra, Lafrouda, Lagos (Lag), Palaia Koiti (PK) and Vistonis (Vis1, Vis2, Vis3) lagoons. The sampling sites are indicated with a red symbol.

Table 1. Latitude, longitude, mean annual conductivity (μS cm^{-1}) and mean annual salinity (psu) per sampling site.

Sampling Site	Lagoon	Latitude	Longitude	Mean Annual Conductivity (μS cm^{-1})	Mean Annual Salinity (psu)
Lafra	Lafra	40°59′32.08″ N	25° 2′29.49″ E	4,3334.9	28.2
Lafrouda	Lafrouda	40°59′11.98″ N	25° 1′51.57″ E	4,7201.4	31.1
Lag	Lagos	40°59′36.13″ N	25° 8′43.93″ E	4,9274.1	29.8
PK	Palaia Koiti	40°54′23.90″ N	24°52′24.32″ E	4,9468.2	30.7
Vis1	Vistonis	41°0′3.58″ N	25° 9′20.89″ E	1,1664.6	7.3
Vis2	Vistonis	41°0′27.24″ N	25° 8′15.77″ E	2,4811.4	13.9
Vis3	Vistonis	41°0′19.28″ N	25° 8′16.10″ E	3,0070.9	18.2

During all samplings, in situ measurements of water temperature and conductivity were made with the use of a HACH HQ40D portable multi meter. Conductivity was transformed to salinity based on the equation in Weyl [36]. Phytoplankton samples of 0.5 L were collected from the surface layer (0–1 m) and were preserved immediately with Lugol's solution. The samples were kept in the dark until they were microscopically analyzed within the next few days.

2.2. Microscopy Analysis

Phytoplankton samples were examined using an inverted epi-fluorescence microscope (Nikon Eclipse TE 2000-S, Melville, MSA) with phase contrast. Phytoplankton identification was based on taxonomic keys and papers and counting was done using the inverted microscope method following the same protocol described in Mazaris et al. [37]. For biomass estimation, the dimensions of 30 individuals of each abundant species were measured using a digital microscope camera (Nikon DS-L1). Mean cell volume estimates were calculated using the appropriate geometric formulae [38]. Phytoplankton biovolume for individuals > 180 μm^3 was converted into carbon content using the formulae $C = 0.288\ V^{0.811}$ for diatoms and $C = 0.216\ V^{0.939}$ for other phytoplankton taxa (C is carbon content in pg, V is cell volume in μm^3) [39]. For individuals < 180 μm^3, conversion factors of 0.108 pg C μm^{-3} for diatoms and 0.157 pg C μm^{-3} for other phytoplankton taxa were used [40].

2.3. Diversity

Alpha diversity estimators for the phytoplankton community (the Shannon and Evenness indices) were calculated with the PAST3 software [41].

To compute beta diversity of the phytoplankton community for each sampling, we used the 'betapart' R package version 1.5.1 [31]. Baselga's [31,42] approach suggests that Sorensen multiple-site dissimilarity (bSOR) is partitioned into two components: spatial turnover in species composition, measured as Simpson dissimilarity (bSIM) and variation in species composition due to nestedness (bNES) measured as nestedness-resultant fraction of Sorensen dissimilarity. Furthermore, Sorensen pair-wise dissimilarity (bSOR) between site pairs was computed in order to use the values for the construction of non-linear regression fitting models. The above analyses were run in R environment version 3.5.3 [43].

2.4. Non-Linear Regression Models

To explore the relationship between conductivity and phytoplankton metrics (abundance, biomass, phytoplankton C, species richness, Evenness and Shannon index) we created non-linear regression models using conductivity as the predictor variable and phytoplankton metrics as the response variable. Models were fit to the data according to the polynomial equation $y = \alpha 1X^2 + a2X + a3$, where y is the response variable, X is the predictor variable and $\alpha 1$, $\alpha 2$ and $\alpha 3$ the model coefficients estimated by maximum likelihood. The statistical significance of the coefficients in the constructed models was assessed by their individual p-value; a p-value < 0.05 indicated a significant coefficient for the predictor variable. The overall significance of each constructed model was assessed by the F-statistic and its p-value. Finally, the strength of the relationship between the predictor and the response variable was determined through the multiple R^2. All the included data for the construction of the models were log(x + 1) transformed in order to produce the smallest error possible when making a prediction and improve the fit of the model [44].

To explore the relationship of beta diversity with conductivity we constructed a model using the difference in conductivity (absolute values) between pair of sites and Sorensen pair-wise dissimilarity (bSOR). The polynomial equation was also used to fit the data, the statistical significance of the coefficients was evaluated by their p-value and the overall significance of the model by the F-statistic and its p-value. Multiple R^2 was used to examine the strength of the relationship between the difference in conductivity and pair-wise bSOR.

2.5. Statistical Analysis

Multidimensional scaling (MDS) based on Jaccard similarity coefficient of presence-absence data was used to depict the similarities of phytoplankton communities among the different lagoons. Significant differences in phytoplankton community structure among the samples were evaluated with one-way analysis of similarities (ANOSIM) and similarity percentage (SIMPER) analysis was run to determine the percentage contribution of phytoplankton species in dissimilarity between the site pairs. Finally, we used a one-way permutational multivariate analysis of variance (PERMANOVA) to test the significance of conductivity on the phytoplankton community composition and biomass among the different sampling sites. The above statistical analyses were run in PAST 3.

3. Results

3.1. Phytoplankton Community Structure and Biomass

A total of 77 phytoplankton species were identified in all the sampling sites during one year (November 2018 to October 2019) which were affiliated to 10 taxonomic groups (Supplementary Table S1). Bacillariophyceae (diatoms) had the highest overall species richness among the taxonomic groups with 31 representatives, followed by Dinophyceae (17 species), Chlorophyceae (10 species) and Cyanobacteria (8 species) (Supplementary Table S1). The rest of the taxonomic groups (Cryptophyceae, Haptophyceae, Prymnesiophyceae, Prasinophyceae, Euglenophyceae & Raphidophyceae) contributed with ≤4 species each to total phytoplankton richness. Diatoms were also the most diverse group in terms of species richness in all the sampling sites. Palaia Koiti lagoon with mean annual conductivity 49,468.2 μS cm^{-1} (mean annual salinity 30.7 psu; Table 1) exhibited the highest diatom richness (28) while Vis1 in Vistonis lagoon with mean annual conductivity 11,664.6 μS cm^{-1} (mean annual salinity 7.3 psu) the lowest (13) (Table 2). Dinoflagellates had higher richness in Lafra, Lafrouda, Lagos and Palaia Koiti lagoons with a mean annual conductivity > 40,000 μS cm^{-1} (mean annual salinity > 28 psu). Chlorophytes and cyanobacteria appeared to be more diverse in Vistonis lagoon with mean annual conductivity ≤ 30,000 μS cm^{-1} (mean annual salinity < 18.5 psu).

Table 2. Number of species for each phytoplankton taxonomic group in Lafra, Lafrouda, Lagos (Lag), Palaia Koiti (PK) and Vistonis (Vis1, Vis2, Vis3) lagoons.

	Lafra	Lafrouda	Lagos	Palaia Koiti	Vis1	Vis2	Vis3
Cyanobacteria	5	5	7	4	8	7	8
Bacillariophyceae	18	21	16	28	13	18	20
Dinophyceae	12	10	15	13	9	8	12
Chlorophyceae	4	2	4	3	8	5	7
Others	9	8	10	9	8	9	8

Phytoplankton biomass varied highly throughout the year in all the sampling sites with the highest biomass values being most commonly measured during spring and summer and the lowest during winter (Figure 2). The highest variability was recorded in Vis1 where the maximum biomass (134.7 mg L^{-1}) was measured in September 2019. This high biomass was made up by the diatom *Rhizosolenia setigera* forming a bloom (12,315 cell mL^{-1}) at a conductivity of 16,090 μS cm^{-1} (9.7 psu). The minimum biomass (4.4 mg L^{-1}) in Vis1 was measured in January 2019 dominated by the diatom *Cyclostephanos* sp. at a conductivity of 2620 μS cm^{-1} (2.6 psu). Similarly, high variability in phytoplankton biomass was recorded in Lafrouda lagoon reaching a maximum biomass of 127.8 mg L^{-1} in April 2019, concurrent with a *Prorocentrum minimum* bloom (301 cell mL^{-1}) at conductivity 46,400 μS cm^{-1} (31.6 psu). The minimum biomass (0.2 mg L^{-1}) in this lagoon was measured in December 2018 and was mainly made up by *Cyclostephanos* sp. at a conductivity of 25,700 μS cm^{-1} (8.6 psu).

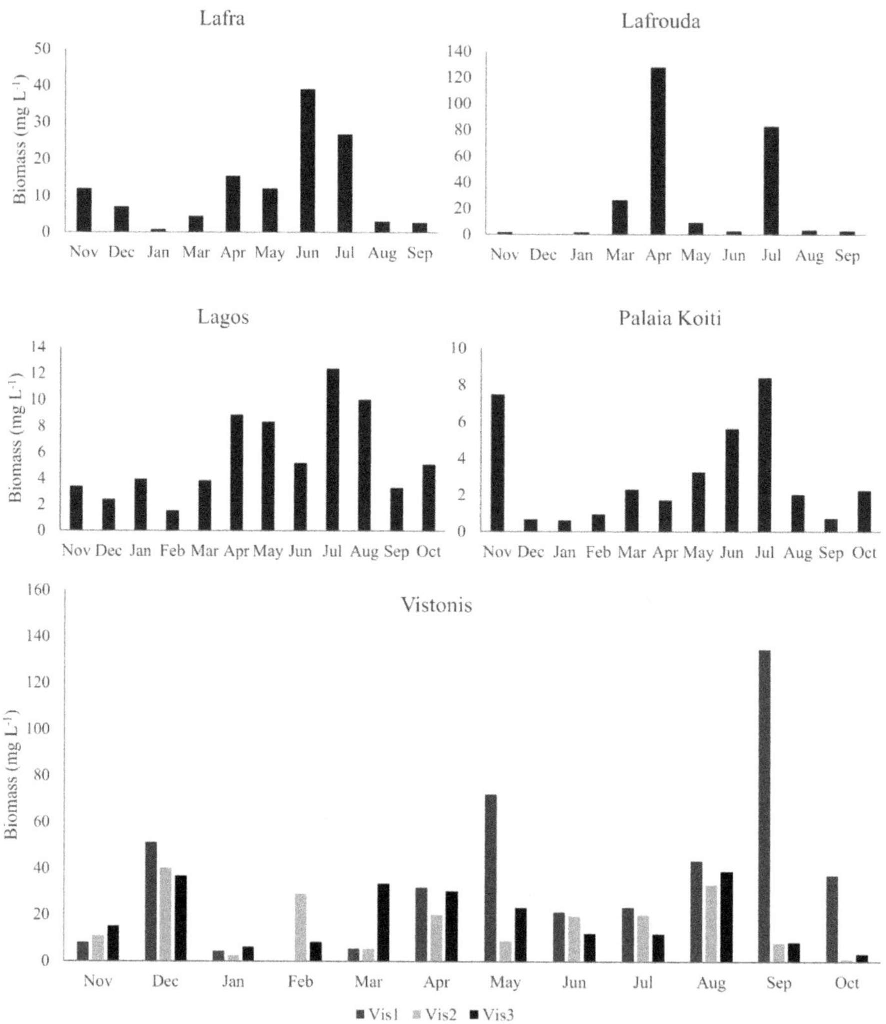

Figure 2. Total phytoplankton biomass in Lafra, Lafrouda, Lagos, Palaia Koiti and Vistonis lagoons during the period from November 2018 to October 2019.

A comparable high variability was noticed for the phytoplankton biomass proxy (phytoplankton C) in all the sampling sites (Supplementary Figure S1a). The widest range was observed in Lafrouda lagoon with a maximum value of 16,448.5 μg C L^{-1} coinciding with the highest biomass (April 2019) and a minimum of 16.9 μg C L^{-1} coinciding with the lowest biomass (December 2018). Vis3 followed with a similar wide range and particularly a maximum value of 12,851.4 μg C L^{-1} was recorded in March 2019 at a conductivity of 23,900 μS cm^{-1} (14.6 psu). The minimum value of phytoplankton C (417.6 μg C L^{-1}) in Vis3 was observed in October 2019 at a conductivity of 31,600 μS cm^{-1} (18.6 psu).

3.2. *Phytoplankton Diversity*

3.2.1. Alpha Diversity Estimators

The number of phytoplankton taxa varied among samples between 7 in December 2018 in Lafrouda lagoon at a conductivity of 25,700 μS cm^{-1} (8.6 psu) and 29 in June 2018 in Lagos lagoon at a conductivity of 50,600 μS cm^{-1} (32.7 psu) (Supplementary Table S2). We found an overall increasing trend in the total species richness over spring and summer in all the sampling sites with the lowest number of taxa being most commonly recorded during winter. The highest number of phytoplankton taxa as well as the highest mean annual number of taxa were found in Lagos (29 taxa) and Palaia Koiti (27 taxa) while the lowest in Lafrouda (7 taxa) (Supplementary Figure S1b). Likewise, the Shannon diversity index showed an increasing trend over spring and summer compared to winter months. The highest values along with the highest mean annual values were computed for Lagos and Palaia Koiti (Supplementary Figure S1c). Evenness remained relatively low throughout the study period in all the sampling sites reflecting significant dominance by few taxa (Supplementary Figure S1d). Hence, the lowest values of Evenness were observed at the same time as heavy phytoplankton blooms (e.g., the case of *Prorocentrum minimum* bloom in April 2019 in Lafrouda lagoon).

3.2.2. Beta Diversity

Phytoplankton beta diversity ranged from 0.6 (January and October 2019) to 0.76 (December 2018) and did not display a seasonal pattern similar to the majority of alpha diversity estimators (e.g., Species richness and Shannon index). Community compositional variation was almost entirely attributed to species turnover (0.60 ± 0.05 SE) rather than nestedness (0.07 ± 0.05 SE; Figure 3). Both turnover and nestedness did not show high fluctuation under different conductivity levels.

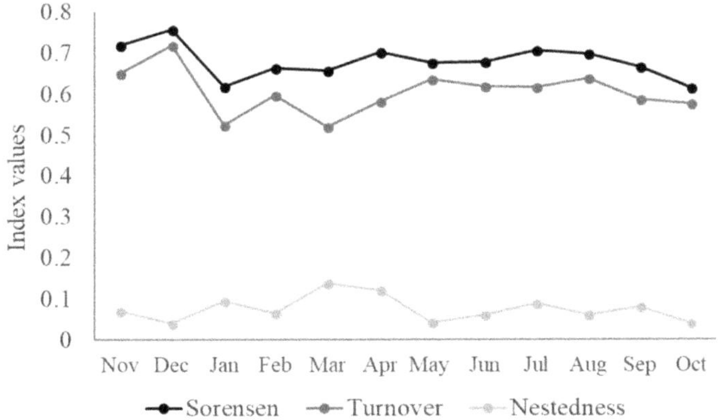

Figure 3. Phytoplankton beta diversity according to Sorensen similarity index (bSOR), turnover (bSIM) and nestedness (bNES) components values in Lafra, Lafrouda, Lagos, Palaia Koiti and Vistonis lagoons during the period from November 2018 to October 2019.

3.3. Phytoplankton Grouping

MDS showed a separation of the phytoplankton community composition based on mean annual conductivity (Figure 4). The majority of the samples from Vistonis lagoon, with a mean annual conductivity ≤ 30,000 μS cm^{-1} (mean annual salinity < 18.5 psu), were placed together and separately from the samples of Lafra, Lafrouda, PK and Lag sampling sites with mean annual conductivity > 40,000 μS cm^{-1} (mean annual salinity > 28 psu). This separation was supported by one-way PERMANOVA which detected significant effects of conductivity ($F = 2.0, p < 0.01$) on the phytoplankton community structure. One-way ANOSIM also indicated significant differences ($R = 0.2, p_{same} < 0.01$) since higher similarities in the phytoplankton community composition were found within the same site in two different time points than between two different sites. According to SIMPER analysis, four to nine species were responsible for > 70% of the compositional differences between the three sampling sites of Vistonis lagoon (Vis1, Vis2, Vis3) and Lafra, Lafrouda, PK and Lag with *Prorcentrum minimum* contributing with > 20% to the compositional differences in all cases (Supplementary Table S3). The other species discriminating the phytoplankton communities among the sampling sites were the diatoms *Chaetoceros* sp., *Cyclostephanos* sp., *Rhizosolenia setigera* and *Skeletonema* cf. *costatum*, the dinoflagellates *Alexandrium* sp. and *Gonyaulax verior* as well as the coccoid cyanobacteria.

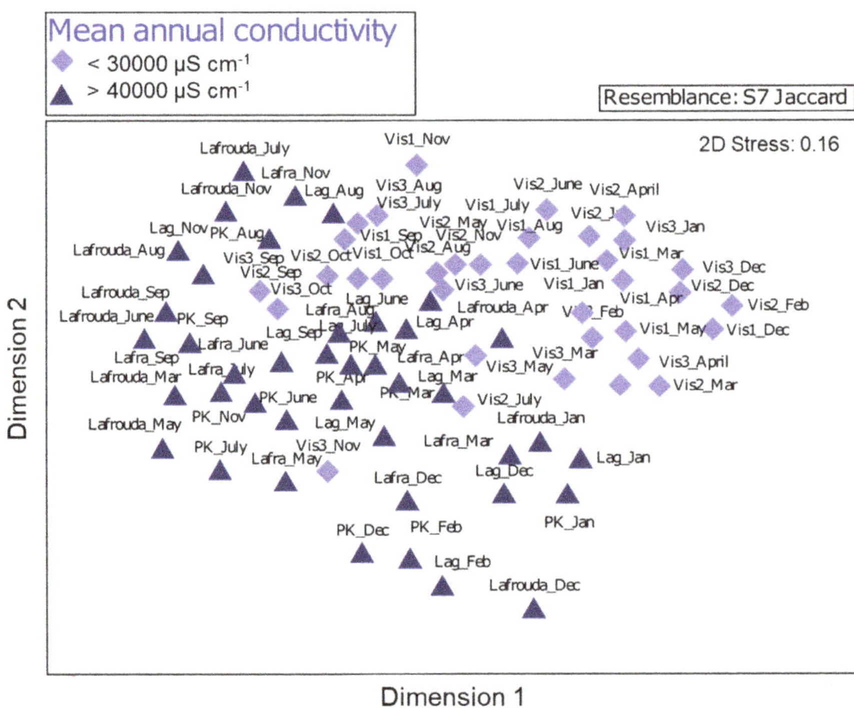

Figure 4. Multidimensional scaling plot of phytoplankton biomass variation in Lafra, Lafrouda, Lagos (Lag), Palaia Koiti (PK) and Vistonis (Vis1, Vis2, Vis3) lagoons according to Jaccard similarity index. Samples with mean annual conductivity lower than 30,000 μS cm^{-1} are represented with a light blue rhomb and samples with mean annual conductivity higher than 40,000 μS cm^{-1} are represented with a dark blue triangle. 2D Stress value: 0.16.

Regarding the biomass values of the discriminating species, *Prorocentrum minimum* reached the extremely high biomass of 124.6 mg L^{-1} in April 2019 in Lafrouda lagoon with a conductivity of 46,400 μS cm^{-1} (31.6 psu). Occasionally, very high biomass was measured for this species in the three sampling sites of Vistonis lagoon; 55.6 mg L^{-1} in May 2019 in Vis1 at conductivity 15,910 μS cm^{-1} (9.6 psu) and 26.1 mg L^{-1} in August 2019 in Vis3 at conductivity 34,900 μS cm^{-1} (22 psu). On the contrary, the biomass of *Prorocentrum minimum* remained rather low (<2.5 mg L^{-1}) in Lagos and Palaia Koiti lagoons with mean annual conductivity > 49,000 μS cm^{-1} (>29 psu). In contrast to *Prorocentrum minimum*, *Alexandrium* sp. and *Gonyaulax verior* were detected only in one or two samples of Vistonis lagoon throughout the study period. However, they were recorded more frequently and with higher biomass in Palaia Koiti, Lagos, Lafra and Lafrouda lagoons. The maximum biomass of *Alexandrium* sp. (81.4 mg L^{-1}) was measured in July 2019 in Lafrouda at a conductivity of 47,400 μS cm^{-1} (30.9 psu) and of *Gonyaulax verior* (3.7 mg L^{-1}) at the same month in Lafra at a conductivity of 36,800 μS cm^{-1} (25.5 psu). Even though the diatoms *Chaetoceros* sp., *Cyclostephanos* sp., *Rhizosolenia setigera* and *Skeletonema* cf. *costatum* were detected in the majority of the samples in the seven sampling sites, they all reached their maximum biomass in Vistonis lagoon, except for *Skeletonema* cf. *costatum*, which had higher biomass in Lagos and Palaia Koiti lagoons. In particular, the maximum biomass of *Chaetoceros* sp. (9.3 mg L^{-1}) was measured in November 2018 in Vis3 at a conductivity of 44,850 μS cm^{-1} (17.2 psu), of *Cyclostephanos* sp. (20.5 mg L^{-1}) in December 2018 in Vis2 at a conductivity of 17,457 μS cm^{-1} (1.7 psu), and of *Rhizosolenia setigera* (123.2 mg L^{-1}) in September 2019 in Vis1 at a conductivity of 16,090 μS cm^{-1} (9.7 psu). For *Skeletonema* cf. *costatum* the maximum biomass (1.4 mg L^{-1}) was measured in August 2019 in PK at a conductivity of 49,300 μS cm^{-1} (26.4 psu). Regarding the coccoid cyanobacteria, they were also detected in the majority of the samples. Nevertheless, their biomass was higher in Vistonis lagoon where they reached their maximum value (6.7 mg L^{-1}) in May 2019 in Vis1 at a conductivity of 15,910 μS cm^{-1} (9.6 psu). Furthermore, in Vistonis lagoon, we measured the maximum biomass for other two cyanobacterial species *Dolichospermum flos-aquae* in September 2019 at a conductivity of 16,090 μS cm^{-1} (9.7 psu) and *Aphanizomenon favalaroi* in November 2018 at a conductivity of 11,716 μS cm^{-1} (12.8 psu). It is worth noting that *Dolichospermum flos-aquae* and *Aphanizomenon favalaroi* were not detected in the samples from the other four lagoons (Lafra, Lafrouda, Lagos, Palaia Koiti) with the exception of two samples in Lagos lagoon where only few individuals were found.

3.4. Non-linear Regression Models

Five out of the seven constructed models demonstrated significant relationships (p-value < 0.05) between conductivity (predictor) and phytoplankton variables (phytoplankton abundance and biomass, phytoplankton C, species richness and pair-wise bSOR) (Table 3). A negative relationship was noticed in all five cases between the predictor and responses variables and in particular, we found that a potential increase in conductivity will lead to significant negative effects on phytoplankton abundance and biomass, phytoplankton C, species richness and pair-wise bSOR (Supplementary Figure S2). The strongest relationship (R^2 of 0.4) was detected between conductivity and phytoplankton abundance while the weakest ($R^2 = 0.1$) was between conductivity and pair-wise bSOR. The lowest residual standard error (0.1) was computed for the constructed model between conductivity and species richness since the majority of our observation data were placed above or very close to the best fit (Supplementary Figure S2).

Table 3. Statistics of the constructed non-linear regression models. Significant relationships (p-value < 0.05) are in bold.

Response Variable	Predictor	Multiple R^2	F-Statistic	*dF*	*p*-Value
Phytoplankton abundance	Conductivity	**0.400**	**22.80**	67	**3.810^{-8}**
Phytoplankton biomass	Conductivity	**0.208**	**8.80**	67	**0.0004**
Phytoplankton C	Conductivity	**0.269**	**12.30**	67	**2.764^{-5}**
Phytoplankton species richness	Conductivity	**0.187**	**7.70**	67	**0.943^{-8}**
Shannon	Conductivity	0.039	1.35	67	0.270
Evenness	Conductivity	0.100	3.60	67	0.053
Pair-wise bSOR	Difference in conductivity	**0.100**	**73.69**	2412	**2.210^{-16}**

4. Discussion

The significant role of salinity/conductivity in determining the composition and diversity of microbial communities in aquatic ecosystems is supported by a series of studies carried out in coastal lagoons (e.g., [45]), in marine environments (e.g., [12]) as well as in lakes (e.g., [46]). Hydrological connectivity has also been identified as an emergent driver for species homogenization between two or more adjacent sites [47]. After discussing how conductivity affected the phytoplankton community structure and diversity across the five studied lagoons, we will discuss the degree of homogeneity (beta diversity) among the communities. Thus, we address the question of whether environmental heterogeneity in terms of conductivity or hydrological connectivity is the primary driver for shaping the less studied phytoplankton communities in transitional water bodies.

4.1. Phytoplankton Composition and Alpha Diversity

In our study, diatoms were the most diverse group with the highest species number in all five lagoons. This result corroborates previous findings on phytoplankton community structure in other Mediterranean lagoons where diatoms supported the highest taxonomic richness [22,24]. Commonly, diatoms are considered as an important component of phytoplankton communities in estuaries and coastal waters being typically the dominant taxonomic group in terms of species richness in such environments [48]. This success and dominance may be attributed to the higher inherent plasticity that several diatom genera (such as *Skeletonema*, *Chaetoceros*, *Thalassiosira*) have in comparison to other phytoplanktonic genera such as the dinoflagellates *Alexandrium* and *Gonyaulax* or the cyanobacterium *Aphanizomenon* and the chlorophyte *Monoraphidium*. Bussard et al. [49] associated the phenotypic plasticity of diatoms under different environmental conditions (including a salinity gradient) with their ability to change their gene expression pathways. They mentioned that this taxonomic group induces the expression of stress-related genes under stress environmental conditions. This characteristic justifies the higher acclimation capability of diatoms to environmental fluctuations in comparison to other phytoplankton taxonomic groups. On the other hand, dinoflagellates due to low inherent plasticity together with higher marine affinity were more diverse in the lagoons with relatively high mean annual conductivity/salinity (> 40,000 $\mu S\ cm^{-1}$ and > 28 psu). This is in contrast to chlorophytes and cyanobacteria which are characterized by higher freshwater affinity in brackish waters [50] and appeared to be more diverse in Vistonis lagoon (< 30,000 $\mu S\ cm^{-1}$ and < 18.5 psu).

Conductivity significantly determined the phytoplankton composition and biomass according to one-way PERMANOVA in line with previous studies in similar ecosystems (e.g., [51,52]). In addition, there was very high variation in the values of the two alpha diversity indices with annual mean values of the Shannon index being double in Palaia Koiti lagoon compared to Lafrouda lagoon. Conductivity did not account or only weakly (non-significantly) accounted for differences in alpha diversity indices across the sampling sites (based on the relationships as demonstrated in the constructed non-linear regression models). Thus, most probably, the different typological descriptors (size, morphometry) of the lagoons were responsible for the high variation in species richness. Smith et al. [53] analyzed data from 142 different natural ponds, lakes, and oceans and 239 experimental ecosystems that revealed a strong phytoplankton species–area relationship. Although there is broad agreement on the positive effect of diversity on community productivity [54,55], during our survey, the lowest values of alpha diversity indices were concurrent with the highest phytoplankton biomass. These findings are well in line with the humped-back biomass–species richness relationship in phytoplankton communities due to the habitat and trophic state of the ecosystem [56]. The resultant biomass–species richness relationship indicates that heavily eutrophic environments as our studied lagoons are expected to have strong dominance and not the optimum species richness and diversity. According to Dodson et al. [57] an optimum of phytoplankton species richness is found at intermediate productivity for lake areas <100 km^2. In the majority of the cases where very high phytoplankton biomass values was measured, it was attributed to the very high proliferation of one or two species (i.e., *Prorocentrum minimum* contributed with 97.5% to the total phytoplankton biomass of 127.8 mg L^{-1} at Lafrouda lagoon measured on April 2019).

Prorocentrum minimum formed blooms in all studied lagoons. This is a harmful, common bloom-forming species in eutrophic, low to moderate salinity coastal and brackish environments [58]. In the Mediterranean region, this species can proliferate in a wide range of water temperatures (4–27 °C) showing great ability to adapt to widely varying natural conditions as shown in our study. *Alexandrium*, also formed blooms at high biomass at different salinities. The ability of *Alexandrium* to colonize multiple habitats and its adaptability and persistence over large region has been recognized in association with the ability of its species to produce toxins [59]. Blooms of the other euryhaline species, the mucilage producing diatoms *Cylindrotheca closterium*, *Rhizosolenia setigera* and *Chaetoceros* sp. have been observed in a eutrophic coastal sea contributing to frequent mucilage events [60].

4.2. High Variation in Conductivity Promotes High Phytoplankton Beta Diversity

Beta diversity is considered positively dependent on environmental heterogeneity since an increase in the latter incorporates an increase in the variety of environmental conditions to which different species are adapted, hence promoting greater variation in species composition among local communities within a region [61,62]. Studies on phytoplankton communities from different aquatic environments have provided consistent evidence and relate the enhanced beta diversity to environmental gradients within a region [63,64]. Under different circumstances, environmental homogeneity has been shown to generate biotic homogenization and thus, low beta diversity of biological communities, including phytoplankton [65]. In aquatic ecosystems, hydrological connectivity has been associated with environmental homogeneity and hence, with species dispersal and high shared diversity throughout a region [66–68]. Even though hydrological connectivity is supported for the five lagoons connected to Vistonikos Gulf by tidal canals, according to our analyses, instead of finding evidence of biotic homogenization (i.e., low beta diversity and low turnover), we calculated relatively high beta diversity for the entire monitoring period. Community compositional variation was almost entirely attributed to species turnover (i.e., species replacement among the different sampling sites). This result contradicts our hypothesis for biotic homogenization and decreased beta diversity, as a consequence of hydrological connectivity of the lagoons.

We assume that conductivity contributed to the species replacement (turnover) between the sampling sites since a significant effect was detected between conductivity and the pair-wise beta diversity (according to the constructed non-linear regression models). Hence, our hypothesis for heterogeneous phytoplankton communities is clearly supported, reflecting the direct effects of environmental heterogeneity, by means of conductivity in our case, on species turnover according to Soininen et al. [69]. It seems that lagoon typological factors (size, morphometry and especially the share of water volume or surface area in contact to the sediment) and local environmental variables were significant determinants of the community structure in agreement with previous studies [62,70]. Our study provides evidence that habitat fragmentation together with environmental heterogeneity constitute an important part of the theory of metacommunity organization [71].

Understanding the relationships between biological communities and environmental aspects remains a challenge for climate research [72] and at the same time, it is essential for establishing water management policies [70]. Therefore, during this study, we modeled conductivity with several phytoplankton metrics in order to determine the type of relationship (negative or positive) between them. As demonstrated by the constructed non-linear models, all phytoplankton metrics (phytoplankton C, abundance, biomass, species richness and pair-wise beta diversity) consistently exhibited a negative relationship with conductivity, implying that a potential increase in the latter will negatively affect them. Nevertheless, this is a first attempt to predict the phytoplankton response of five Mediterranean lagoons to enhanced salinity due to climate change and longer time series in the future will allow us to strengthen the accuracy of the model and of projections [18].

5. Conclusions

Conductivity/salinity was a key environmental factor in shaping phytoplankton communities across five Mediterranean lagoons connected to the Vistonikos Gulf. In particular, we showed that environmental heterogeneity, by means of conductivity, led to high phytoplankton beta diversity due to high turnover exceeding the effects of hydrological connectivity. This fact implies that local conditions in each lagoon were important in determining which species were capable not only of maintaining a viable population but also of forming a heavy bloom. Nevertheless, the exposure of phytoplankton to continuous salinity fluctuations had as a consequence the increase of euryhaline harmful species. Overall, our results provide useful information for ecological water management of transitional eutrophic water bodies in the Mediterranean region.

Supplementary Materials: The following are available online at http://www.mdpi.com/1424-2818/12/1/38/s1. Supplementary Figure S1: Boxplots of phytoplankton C (a), species richness (b), Shannon index (c) and Evenness (d) in the lagoons Lafra, Lafrouda, Lagos (Lag), Palaia Koiti (PK) and Vistonis (Vis1, Vis2, Vis3). Supplementary Figure S2: Scatter plots conductivity and the different phytoplankton metrics. Supplementary Table S1: Species list of phytoplankton taxa per sampling site during the study period. Supplementary Table S2: Species number, total phytoplankton biomass and alpha diversity indices per sample. Supplementary Table S3: List of higher contributing phytoplankton species in communities' dissimilarity determined by SIMPER between site pairs.

Author Contributions: Conceptualization, N.S. and M.M.-G.; Sampling D.T., Microscopy Analysis, N.S., M.K.; Data analysis N.S.; Resources, M.M.-G.; Writing-Original Draft Preparation, N.S. and M.M.-G.; Writing Review and Editing N.S., M.M.-G., M.K., E.M., M.D., D.T.; Visualization, N.S., M.D.; Supervision, M.M.-G.; Project Administration, M.K., D.T. and M.M.-G.; Funding Acquisition, M.M.-G. All authors have read and agreed to the published version of the manuscript.

Funding: This research was partially funded by PREFECTURE OF EASTERN MACEDONIA AND THRACE, grant number ELKE-AUTh: 97071.

Conflicts of Interest: The authors declare no conflict of interest.

References

1. Kjerfve, B. *Coastal Lagoon Processes*; Elsevier Science Publishers: Amsterdam, The Netherland, 1994; p. 577.
2. Knoppers, B. Aquatic primary production in coastal lagoons. In *Coastal Lagoon Processes*; Kjerfve, B., Ed.; Elsevier Science Publishers: Amsterdam, The Netherland, 1994; pp. 243–286.
3. Levin, L.A.; Boesch, D.F.; Covich, A.; Dahm, C.; Erséus, C.; Ewel, K.C.; Kneib, R.T.; Moldenke, A.; Palmer, M.A.; Snelgrove, P.; et al. The function of marine critical transition zones and the importance of sediment biodiversity. *Ecosystems* **2000**, *4*, 430–451. [CrossRef]
4. Mouillot, D.; Gaillard, S.; Aliaume, C.; Verlaque, M.; Belsher, T.; Troussellier, M.; Thang, D.C. Ability of taxonomic diversity indices to discriminate coastal lagoon environments based on macrophyte communities. *Ecol. Indic.* **2005**, *5*, 1–17. [CrossRef]
5. Lloret, J.; Marín, A.; Marín-Guirao, L. Is coastal lagoon eutrophication likely to be aggravated by global climate change? *Estuar. Coast. Shelf Sci.* **2008**, *78*, 403–412. [CrossRef]
6. De Casabianca, M.L.; Laugier, T.; Marinho-Soriano, E. Seasonal changes of nutrients in water and sediment in a Mediterranean lagoon with shellfish farming activity (Thau Lagoon, France). *ICES Mar. Sci.* **1997**, *54*, 905–916. [CrossRef]
7. Pérez-Ruzafa, A.; Fernández, A.I.; Concepción, M.; Gilabert, J.; Quispe, I.J.; García-Charton, J.A. Spatial and temporal variations of hydrological conditions, nutrients and chlorophyll a in a Mediterranean coastal lagoon (Mar Menor, Spain). *Hydrobiologia* **2005**, *550*, 11–27. [CrossRef]
8. Grigorakis, K.; Rigos, G. Aquaculture effects on environmental and public welfare—The case of Mediterranean mariculture. *Chemosphere* **2011**, *85*, 899–919. [CrossRef]
9. Eisenreich, S.J. (Ed.) *Climate Change and the European Water Dimension*; EU-Report 21553 of the European Commission; Joint Research Centre: Ispra, Italy, 2005; p. 253.
10. Intergovernmental Panel on Climate Change [IPCC]. *Climate Change 2014: Impacts, Adaptation and Vulnerability*; IPCC Working Group II Contribution to the 5th Assessment Report of the International Panel on Climate Change; Cambridge University Press: Cambridge, UK, 2014; p. 169.

11. Intergovernmental Panel on Climate Change [IPCC]. Summary for Policymakers. In *Global Warming of 1.5 °C; An IPCC special report on the impacts of global warming of 1.5 °C above pre-industrial levels and related global greenhouse gas emission pathways, in the context of strengthening the global response to the threat of climate change, sustainable development, and efforts to eradicate poverty*; World Meteorological Organization: Geneva, Switzerland, 2018; p. 630.
12. Stefanidou, N.; Genitsaris, S.; Lopez-Bautista, J.; Sommer, U.; Moustaka-Gouni, M. Effects of heat shock and salinity changes on coastal Mediterranean phytoplankton in a mesocosm experiment. *Mar. Biol.* **2018**, *165*, 154. [CrossRef]
13. Stefanidou, N.; Genitsaris, S.; Lopez-Bautista, J.; Sommer, U.; Moustaka-Gouni, M. Unicellular eukaryotic community response to temperature and salinity variation in mesocosm experiments. *Front. Microbiol.* **2018**, *9*, 2444. [CrossRef]
14. Pérez-Ruzafa, A.; Marcos, C.; Pérez-Ruzafa, I.M. Mediterranean coastal lagoons in an ecosystem and aquatic resources management context. *Phys. Chem. Earth Parts A B C* **2011**, *36*, 160–166. [CrossRef]
15. Blanchet, M.; Pringault, O.; Bouvy, M.; Catala, P.; Oriol, L.; Caparros, J.; Ortega-Retuerta, E.; Intertaglia, L.; West, N.; Agis, M.; et al. Changes in bacterial community metabolism and composition during the degradation of dissolved organic matter from the jellyfish *Aurelia aurita* in a Mediterranean coastal lagoon. Environ. *Sci. Pollut. Res.* **2015**, *22*, 13638. [CrossRef]
16. Ghai, R.; Hernandez, C.M.; Picazo, A.; Mizuno, C.M.; Ininbergs, K.; Diez, B.; Valas, R.; DuPont, C.L.; McMahon, K.D.; Camacho, A.; et al. Metagenomes of Mediterranean coastal lagoons. *Sci. Rep.* **2012**, *2*, 490. [CrossRef] [PubMed]
17. Riera, R.; Tuset, V.M.; Betancur-R, R.; Lombarte, A.; Marcos, C.; Pérez-Ruzafa, A. Modelling alpha-diversities of coastal lagoon fish assemblages from the Mediterranean Sea. *Prog. Oceanogr.* **2018**, *165*, 100–109. [CrossRef]
18. Bellino, A.; Mangano, M.C.; Baldantoni, D.; Russell, B.D.; Mannino, A.M.; Mazzola, A.; Vizzini, A.; Sarà, G. Seasonal patterns of biodiversity in Mediterranean coastal lagoons. *Divers. Distrib.* **2019**, *25*, 1512–1526. [CrossRef]
19. Orfanidis, S.; Pinna, M.; Sabetta, L.; Stamatis, N.; Nakou, K. Variation of structural and functional metrics in macrophyte communities within two habitats of eastern Mediterranean coastal lagoons: Natural versus human effects. *Aquat. Conserv. Mar. Freshw. Ecosyst.* **2008**, *18*, S45–S61. [CrossRef]
20. Koutrakis, E.T.; Tsikliras, A.C.; Sinis, A.I. Temporal variability of the ichthyofauna in a Northern Aegean coastal lagoon (Greece). Influence of environmental factors. *Hydrobiologia* **2005**, *543*, 245. [CrossRef]
21. Carlier, A.; Riera, P.; Amouroux, J.M.; Bodiou, J.Y.; Desmalades, M.; Grémare, A. Food web structure of two Mediterranean lagoons under varying degree of eutrophication. *J. Sea Res.* **2008**, *60*, 264–275. [CrossRef]
22. Roselli, L.; Stanca, E.; Ludovisi, A.; Durante, G.; Souza, J.S.D.; Dural, M.; Alp, T.; Bulent, S.; Gjoni, V.; Ghinis, S.; et al. Multi-scale biodiverity patterns in phytoplankton from coastal lagoons: The Eastern Mediterranean. *Transit. Waters Bull.* **2013**, *7*, 202–219.
23. Bazin, P.; Jouenne, F.; Deton-Cabanillas, A.F.; Pérez-Ruzafa, A.; Véron, B. Complex patterns in phytoplankton and microeukaryote diversity along the estuarine continuum. *Hydrobiologia* **2014**, *726*, 155. [CrossRef]
24. Leruste, A.; Pasqualini, V.; Garrido, M.; Malet, N.; De Wit, R.; Bec, B. Physiological and behavioral responses of phytoplankton communities to nutrient availability in a disturbed Mediterranean coastal lagoon. *Estuar. Coast. Shelf Sci.* **2019**, *219*, 176–188. [CrossRef]
25. Trombetta, T.; Vidussi, F.; Mas, S.; Parin, D.; Simier, M.; Mostajir, B. Water temperature drives phytoplankton blooms in coastal waters. *PLoS ONE* **2019**, *14*, e0214933. [CrossRef]
26. Pavloudi, C.; Oulas, A.; Vasileiadou, K.; Sarropoulou, E.; Kotoulas, G.; Arvanitidis, C. Salinity is the major factor influencing the sediment bacterial communities in a Mediterranean lagoonal complex (Amvrakikos Gulf, Ionian Sea). *Mar. Genom.* **2016**, *28*, 71–81. [CrossRef] [PubMed]
27. Grzebyk, D.; Audic, S.; Lasserre, B.; Abadie, E.; de Vargas, C.; Bec, B. Insights into the harmful algal flora in northwestern Mediterranean coastal lagoons revealed by pyrosequencing metabarcodes of the 28S rRNA gene. *Harmful Algae* **2017**, *68*, 1–16. [CrossRef] [PubMed]
28. Aires, T.; Muyzer, G.; Serrão, E.A.; Engelen, A.H. Seaweed Loads Cause Stronger Bacterial Community Shifts in Coastal Lagoon Sediments Than Nutrient Loads. *Front. Microbiol.* **2019**, *9*, 3283. [CrossRef] [PubMed]
29. Whittaker, R.J.; Willis, K.J.; Field, R. Scale and species richness: Towards a general, hierarchical theory of species diversity. *J. Biogeogr.* **2001**, *28*, 453–470. [CrossRef]

30. Koleff, P.; Gaston, K.J.; Lennon, J.J. Measuring beta diversity for presence–absence data. *J. Anim. Ecol.* **2003**, *72*, 367–382. [CrossRef]
31. Baselga, A.; Orme, C.D.L. Betapart: An R package for the study of beta diversity. *Methods Ecol. Evol.* **2012**, *3*, 808–812. [CrossRef]
32. Baselga, A. Partitioning the turnover and nestedness components of beta diversity. *Glob. Ecol. Biogeogr.* **2010**, *19*, 134–143. [CrossRef]
33. Papadakis, E.M.; Tsaboula, A.; Kotopoulou, A.; Kintzikoglou, K.; Vryzas, Z.; Papadopoulou-Mourkidou, E. Pesticides in the surface waters of Lake Vistonis Basin, Greece: Occurrence and environmental risk assessment. *Sci. Total Environ.* **2015**, *536*, 793–802. [CrossRef]
34. Arabatzis, G.D.; Kokkinakis, A.K. Typology of the lagoons of Northern Greece according to their environmental characteristics and fisheries production. *Oper. Res. Int. J.* **2005**, *5*, 21–34. [CrossRef]
35. Koutrakis, E. *Northern Greece Lagoon Management, Suggestions and Problems*; Institute of Plant Breeding and Genetic Resources, Hellenic Agricultural Organization-Inale: Kavala, Greece, 2000; p. 50.
36. Weyl, P.K. On the change in electrical conductance of seawater with temperature. *Limnol. Oceanogr.* **1964**, *9*, 75–78. [CrossRef]
37. Mazaris, A.D.; Moustaka-Gouni, M.; Michaloudi, E.; Bobori, D.C. Biogeographical patterns of freshwater micro- and macroorganisms: A comparison between phytoplankton, zooplankton and fish in the eastern Mediterranean. *J. Biogeogr.* **2010**, *37*, 1341–1351. [CrossRef]
38. Hillebrand, H.; Dürselen, C.-D.; Kirschtel, D.; Pollingher, U.; Zohary, T. Biovolume calculation for pelagic and benthic microalgae. *J. Phycol.* **1999**, *35*, 403–424. [CrossRef]
39. Menden-Deuer, S.; Lessard, E.J. Carbon to volume relationships for dinoflagellates, diatoms, and other protist plankton. *Limnol. Oceanogr.* **2000**, *45*, 569–579. [CrossRef]
40. Sommer, U.; Lengfellner, K.; Lewandowska, A. Experimental induction of a coastal spring bloom early in the year by intermittent high-light episodes. *Mar. Ecol. Prog. Ser.* **2012**, *446*, 61–71. [CrossRef]
41. Hammer, Ø.; Harper, D.A.T.; Ryan, P.D. Past: Paleontological statistics software package for education and data analysis. *Palaeontol. Electron.* **2001**, *4*, 9–10.
42. Baselga, A. Multiplicative partition of true diversity yields independent alpha and beta components; additive partition does not. *Ecology* **2010**, *91*, 1974–1981. [CrossRef]
43. R Core Team. *R: A Language and Environment for Statistical Computing*; R Foundation for Statistical Computing: Vienna, Austria, 2018; Available online: https://www.R-project.org/ (accessed on 20 May 2019).
44. Benoit, K. *Linear Regression Models with Logarithmic Transformations*; London Methodology Institute, London School of Economics: London, UK, 2011; p. 25.
45. Hemraj, D.A.; Hossain, M.A.; Ye, Q.; Qin, J.G.; Leterme, S.C. Plankton bioindicators of environmental conditions in coastal lagoons. *Estuar. Coast. Shelf Sci.* **2017**, *184*, 102–114. [CrossRef]
46. Yang, J.; Ma, L.A.; Jiang, H.; Wu, G.; Dong, H. Salinity shapes microbial diversity and community structure in surface sediments of the Qinghai-Tibetan Lakes. *Sci. Rep.* **2016**, *6*, 25078. [CrossRef]
47. Pérez-Ruzafa, A.; De Pascalis, F.; Ghezzo, M.; Quispe-Becerra, J.L.; Hernández-García, R.; Muñoz, I.; Vergara, C.; Pérez-Ruzafa, I.M.; Umgiesser, G.; Marcos, C. Connectivity between coastal lagoons and sea: Asymmetrical effects on assemblages' and populations' structure. *Estuar. Coast. Shelf Sci.* **2019**, *216*, 171–186. [CrossRef]
48. Carstensen, J.; Klais, R.; Cloern, J.E. Phytoplankton blooms in estuarine and coastal waters: Seasonal patterns and key species. *Estuar. Coast. Shelf Sci.* **2015**, *162*, 98–109. [CrossRef]
49. Bussard, A.; Corre, E.; Hubas, C.; Duvernois-Berthet, E.; Le Corguillé, G.; Jourdren, L.; Coulpier, F.; Claquin, P.; Lopez, P.J. Physiological adjustments and transcriptome reprogramming are involved in the acclimation to salinity gradients in diatoms. *Environ. Microbiol.* **2017**, *19*, 909–925. [CrossRef] [PubMed]
50. Michaloudi, E.; Moustaka-Gouni, M.; Gkelis, S.; Pantelidakis, K. Plankton community structure during an ecosystem disruptive algal bloom of *Prymnesium parvum*. *J. Plankton Res.* **2009**, *31*, 301–309. [CrossRef]
51. Adolf, J.E.; Burns, J.; Walker, J.K.; Gamiao, S. Near shore distributions of phytoplankton and bacteria in relation to submarine groundwater discharge-fed fishponds, Kona coast, Hawaii, USA. *Estuar. Coast. Shelf Sci.* **2019**, *219*, 341–353. [CrossRef]

52. Draredja, M.A.; Frihi, H.; Boualleg, C.; Gofart, A.; Abadie, E.; Laabir, M. Seasonal variations of phytoplankton community in relation to environmental factors in a protected meso-oligotrophic southern Mediterranean marine ecosystem (Mellah lagoon, Algeria) with an emphasis of HAB species. *Environ. Monit. Assess.* **2019**, *191*, 603. [CrossRef] [PubMed]
53. Smith, L.C.; Sheng, Y.; MacDonald, G.M.; Hinzman, L.D. Disappearing Arctic lakes. *Science* **2005**, *308*, 1429. [CrossRef]
54. Stachowicz, J.J.; Best, R.J.; Bracken, M.E.S.; Graham, M.H. Complementarity in marine biodiversity manipulations: Reconciling divergent evidence from field and mesocosm experiments. *Proc. Natl. Acad. Sci. USA* **2008**, *105*, 18842–18847. [CrossRef]
55. Lewandowska, A.M.; Breithaupt, P.; Hillebrand, H.; Hoppe, H.G.; Jürgens, K.; Sommer, U. Responses of primary productivity to increased temperature and phytoplankton diversity. *J. Sea Res.* **2012**, *72*, 87–93. [CrossRef]
56. Török, P.; Krasznai, E.T.; Béres, V.B.; Bácsi, I.; Borics, B.G.; Tóthmérész, B. Functional diversity supports the biomass-diversity humped-back relationship in phytoplankton assemblages. *Funct. Ecol.* **2016**, *30*, 1593–1602. [CrossRef]
57. Dodson, S.I.; Arnott, S.E.; Cottingham, K.L. The relationship in lake communities between primary productivity and species richness. *Ecology* **2000**, *81*, 2662–2679. [CrossRef]
58. Heil, C.A. Influence of humic, fulvic and hydrophilic acids on the growth, photosynthesis and respiration of the dinoflagellate *Prorocentrum minimum* (Pavillard) Schiller. *Harmful Algae* **2005**, *4*, 603–618. [CrossRef]
59. Anderson, D.M. Bloom dynamics of toxic Alexandrium species in the northeastern US. *Limnol. Oceanogr.* **1997**, *42*, 1009–1022. [CrossRef]
60. Genitsaris, S.; Stefanidou, N.; Sommer, U.; Moustaka-Gouni, M. Phytoplankton Blooms, Red Tides and Mucilaginous Aggregates in the Urban Thessaloniki Bay, Eastern Mediterranean. *Diversity* **2019**, *11*, 136. [CrossRef]
61. Tonkin, J.D.; Stoll, S.; Jähnig, S.C.; Haase, P. Contrasting metacommunity structure and beta diversity in an aquatic-floodplain system. *Oikos* **2016**, *125*, 686–697. [CrossRef]
62. Heino, J.; Soininen, J.; Alahuhta, J.; Lappalainen, J.; Virtanen, R. Metacommunity ecology meets biogeography: Effects of geographical region, spatial dynamics and environmental filtering on community structure in aquatic organisms. *Oecologia* **2017**, *183*, 121. [CrossRef]
63. Maloufi, S.; Catherine, A.; Mouillot, D.; Louvard, C.; Couté, A.; Bernard, C.; Troussellier, M. Environmental heterogeneity among lakes promotes hyper β-diversity across phytoplankton communities. *Freshw. Biol.* **2016**, *61*, 633–645. [CrossRef]
64. Chaparro, G.; Horváth, Z.; O'Farrell, I.; Ptacnik, R.; Hein, T. Plankton metacommunities in floodplain wetlands under contrasting hydrological conditions. *Freshw. Biol.* **2018**, *63*, 380–391. [CrossRef]
65. Zhang, Y.; Peng, C.; Huang, S.; Wang, J.; Xiong, X.; Li, D. The relative role of spatial and environmental processes on seasonal variations of phytoplankton beta diversity along different anthropogenic disturbances of subtropical rivers in China. *Environ. Sci. Pollut. Res.* **2019**, *26*, 1422. [CrossRef]
66. Bozelli, R.L.; Thomaz, S.M.; Padial, A.A.; Lopes, P.M.; Bini, L.M. Floods decrease zooplankton beta diversity and environmental heterogeneity in an Amazonian floodplain system. *Hydrobiologia* **2015**, *753*, 233–241. [CrossRef]
67. Gao, Y.; Sassenhagen, I.; Richlen, M.L.; Donald, M.; Anderson, D.M.; Martin, J.L.; Erdner, D.L. Spatiotemporal genetic structure of regional-scale Alexandrium catenella dinoflagellate blooms explained by extensive dispersal and environmental selection. *Harmful Algae* **2019**, *86*, 46–54. [CrossRef]
68. Katsiapi, M.; Genitsaris, S.; Stefanidou, N.; Tsavdaridou, A.; Giannopoulou, I.; Stamou, G.; Michaloudi, E.; Mazaris, A.; Moustaka-Gouni, M. Ecological connectivity in two ancient lakes: Impact upon planktonic cyanobacteria and water quality. *Water* **2020**, *12*, 18. [CrossRef]
69. Soininen, J.; Heino, J.; Wang, J. A meta-analysis of nestedness and turnover components of beta diversity across organisms and ecosystems. *Glob. Ecol. Biogeogr.* **2018**, *27*, 96–109. [CrossRef]
70. Moustaka-Gouni, M.; Sommer, U.; Economou-Amilli, A.; Arhonditsis, G.B.; Katsiapi, M.; Papastergiadou, E.; Kormas, A.K.; Vardaka, E.; Karayanni, H.; Papadimitriou, T. Implementation of the Water Framework Directive: Lessons Learned and Future Perspectives for an Ecologically Meaningful Classification Based on Phytoplankton of the Status of Greek Lakes, Mediterranean Region. *Environ. Manag.* **2019**, *64*, 675. [CrossRef] [PubMed]

71. Leibold, M.A.; Holyoak, M.; Mouquet, N.; Amarasekare, P.; Chase, J.M.; Hoopes, M.F.; Holt, R.D.; Shurin, J.B.; Law, R.; Tilman, D.; et al. The metacommunity concept: A framework for multi-scale community ecology. *Ecol. Lett.* **2004**, *7*, 601–613. [CrossRef]
72. Tunney, T.D.; McCann, K.S.; Lester, N.P.; Brian, J.; Shuter, B.J. Differential warming affects complex communities. *Proc. Natl. Acad. Sci. USA* **2014**, *22*, 8077–8082. [CrossRef]

© 2020 by the authors. Licensee MDPI, Basel, Switzerland. This article is an open access article distributed under the terms and conditions of the Creative Commons Attribution (CC BY) license (http://creativecommons.org/licenses/by/4.0/).

diversity

MDPI

Article

Effect of Hydrographic Variability on the Distribution of Microbial Communities in Taiwan Strait in Winter

Gwo-Ching Gong [1,2], Hsin-Ming Yeh [3], Yu-Kai Chen [3], Chih-hao Hsieh [4], Pei-Chi Ho [4] and An-Yi Tsai [1,*]

1 Institute of Marine Environment and Ecology, National Taiwan Ocean University, Keelung 202-24, Taiwan; gcgong@mail.ntou.edu.tw

2 Center of Excellence for the Oceans, National Taiwan Ocean University Keelung 202-24, Taiwan

3 Marine Fisheries Division, Fisheries Research Institute, Council of Agriculture, Taipei 202, Taiwan; hmyeh@mail.tfrin.gov.tw (H.-M.Y.); ykchen@mail.tfrin.gov.tw (Y.-K.C.)

4 Institute of Oceanography, Nation Taiwan University, Taipei 10617, Taiwan; chsieh@ntu.edu.tw (C.-h.H.); bookwormpageho@gmail.com (P.-C.H.)

* Correspondence: anyitsai@mail.ntou.edu.tw; Tel.: +886-2-2462219; Fax: +886-2-2462-0892

Received: 2 September 2019; Accepted: 11 October 2019; Published: 14 October 2019

Abstract: This study investigated the spatial variation in the components of a microbial food web (viruses, picoplankton, nanoflagellates, and ciliates) in different hydrographic environments in the Taiwan Strait during winter. Water temperature and salinity varied spatially, with lower temperatures (15.3–22.8 °C) and salinities (32.2–33.4 psu) in the northern part of the Taiwan Strait, largely affected by runoff from the coast of China. Concentrations of nutrients and Chl *a* were significantly higher in the northern part than that in the southern part of the study area. *Synechococcus* spp., nanoflagellate, and ciliate abundance also varied significantly, with the northern strait having higher abundances of these communities. In contrast, a higher abundance of bacteria was found in the southern part of the Taiwan Strait. The results of this study, which describes two different ecosystems in the Taiwan Strait, suggest that during winter, a "viral loop" might play an important role in controlling bacterial production in the southern part of the Taiwan Strait, while nanofalgellate grazing of picophytoplankton may contribute mainly to the flux of energy in the northern part.

Keywords: microbial food web; bacteria; nanoflagellates; ciliates; Taiwan Strait

1. Introduction

The microbial community, which consists of bacteria, nanoflagellates, and microzooplankton, plays an important role in the life of aquatic environments [1]. An understanding of the relationship between the distributions of microbial organisms and their environments can help us better understand their eco-physiological requirements. Bacteria are able to use dissolved organic matter and incorporate it into their own biomass. Their biomass may be reduced in several ways, with the important one being grazing by phagotrophic protists (nano- and microzooplankton) and viral lysis [2–5]. These studies showed that both viral lysis and grazing can cause significant mortality, but that the impact of each varies by season, host organism, and environmental conditions. Seasonally, viral lysis is reported to be the main cause of bacterial mortality during the winter, and thus a "viral loop" might serve as an important control mechanism for bacterial production in colder seasons, while grazing can be the cause of most bacterial mortality during warmer seasons [5].

The growth of bacteria can also be controlled by physical factors, such as temperature and the supply of substrates [6–8]. Li [9] reported that their regulation can be seasonal and suggested that, over the course of a year, temperature is the dominant factor affecting bacterial growth in colder seasons. Other factors, such as substrate supply, may be important in warmer seasons. Ochs et al. [10] found that bacterial growth rates are unrelated to temperatures above 14 °C, which was further confirmed by Shiah and Ducklow [11].

Understanding the principles that govern trophic interactions among microbial planktonic organisms, those that regulate carbon and energy flows in microbial food webs, is an important goal in biological oceanography. The dominating predators of bacteria are heterotrophic nanoflagellates (HNFs) [3,12,13]. Some studies reported increases in HNFs in areas where there are increases in bacteria, suggesting that HNFs are resource controlled [12,14]. Other studies have found a weak coupling between bacteria and HNF density and recommended predation control [15,16]. Although correlative data can suggest possible governing factors, such data do not provide much insight into the complex mechanisms involved. For example, the lack of a strong relationship between HNFs and bacterial abundance might be related to the possibility that HNFs use other sources of carbon (e.g., picophytoplanktonic cyanobacteria) [13], or it may be that there are other more important predators of bacteria (e.g., ciliates) [17]. Other studies suggested that bottom-up control dominates HNF abundance [18,19], while other studies noted that top-down control is more important [16,20]. In addition, the relative importance of these controls may change along with the environmental trophic state, season, and other environmental factors [21,22].

Taiwan is located at the junction of Ryukyu and Luzon in the northwestern part of the Pacific Ocean. This large island country is bordered to the west by the shallow Taiwan Strait and to the east by the deep Philippine Sea. The broad (140 to 200 km) Taiwan Strait Shelf is located between the coasts of China and Taiwan and extends for more than 300 km. Generally, currents in the Taiwan region vary consistently from season to season [23]. Jan et al. [24] described the seasonal variations in currents in the Taiwan Strait in detail. During winter, the northward flow of South China Sea waters through the Penghu Channel is slowed down by the northeasterly winter monsoon and turns northwestward along the southern edge of the Changyun Ridge. In the northern area of the Taiwan Strait, the cold and fresh southward flow of China's coastal waters turn cyclonically at the northern edge of the Changyun Ridge, where they form a cyclonic cold eddy. These winter differences in physicochemical and biotic factors in the shallow Taiwan Strait may set the scene for differences in the development of microorganisms and provide an excellent environmental frame to analyze the potential changes in microbial food web components under very different trophic environments. We hypothesize that the spatial variation in temperature affects the microbial dynamics in the shallow Taiwan Strait in cold seasons. Therefore, this study investigated the surface water distributions of viral, heterotrophic bacterial, *Synechococcus* spp., nanoflagellate, and ciliate abundance, as well as possible HNF-bacterial coupling along different cruise transects in the Taiwan Strait during winter.

2. Material and Methods

2.1. Study Sites and Samplings

Samples were collected in January and February 2018 from the surface waters at 25 established stations (9 transects; A–I) located in the Taiwan Strait on board the Fishery Researcher I (Figure 1A). During this cruise, seawater samples were collected at depths of 2 m using a SeaBird CTD (Sea-Bird Scientific, Bellevue, WA, USA) and General Oceanic Rosette assembly with 10 L Niskin bottle (General Oceanic, Florida, USA) Temperature and salinity were measured with a SeaBird CTD. Water samples were filtered (25 mm GF/F: Whatman glass microfiber filters) for Chl *a* analysis and measured after extraction using an in vitro fluorometer (Turner Design 10-AU-005) (Turner Designs, Inc, San Jose, CA, USA) [25]. Nutrients in seawater samples were measured as previously described by Gong et al. [25]. Sample volumes of 50 mL were used to count bacteria, *Synechococcus* spp., and nanoflagellates. They were fixed with gluteraldehyde at a final concentration of 0.5%. For ciliates, 500 mL water samples from the surface were fixed with neutralized formaldehyde (2% final concentration) [26] and preserved at 4 °C until analysis. Virus populations were identified and measured using flow cytometry (FCM) (BD FACSCalibur™) (Biosciences, Macquarie Research Park, Sydney, Australia) [27].

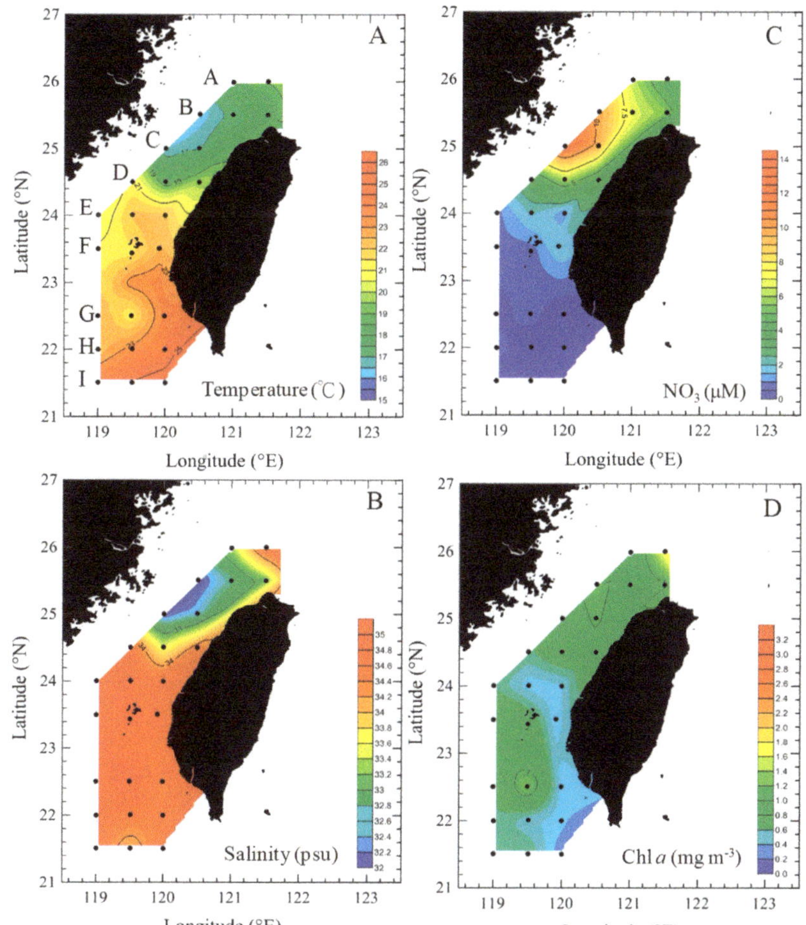

Figure 1. Spatial variations of temperature (**A**), salinity (**B**), NO_3 (**C**), and Chl *a* concentrations (**D**) during the study period in the Taiwan Strait.

2.2. Microbial Counts

Bacteria, *Synechococcus* spp., and nanoflagellates were counted using an epifluorescence microscope (Nikon Optiphot-2) (1000×). Subsamples of 1–2 mL, 4 mL, and 20 mL were filtered onto 0.2 μm or 0.8 μm black Nuclepore filters for bacteria, *Synechococcus* spp., and nanoflagellates, respectively. Samples were stained with DAPI (4′,6-diamidino-2-phenylindole) at a final concentration of 1 μg mL^{-1} [28] to count bacteria and heterotrophic nanoflagellates (HNFs). Pigmented nanoflagellates (PNFs) and HNFs were detected and counted based on the absence or presence of chlorophyll autofluorescence using a separate filter set optimized for chlorophyll or DAPI under a 1000× epifluorescence microscope (Nikon Optiphot-2). Bacteria and HNFs were identified using their blue fluorescence under UV illumination. *Synechococcus* spp. and PNFs were identified using their orange and red autofluorescences under blue excitation light. Furthermore, to obtain reliable ciliate counts, a 500-mL water sample was concentrated into a 100-mL subsample using a 20-μm mesh size net, and then the subsamples (100 mL) were allowed to settle in an Utermöhl chamber (Utermöhl). The entire area of the Utermöhl chamber was examined at 200× or 400× magnification using an inverted microscope (Nikon-TMD 300). To perform FCM analysis, we diluted them to 1:10 with a 0.2-μm filtered TE buffer (Tris-EDTA buffer) (10 mM Tris, 1 mM EDTA: Ethylenediaminetetraacetic acid), stained them with SYBR Green I solution (1:500 dilution; Molecular Probes), and incubated them at 80 °C in the dark for 10 min [29]. Cell side scatter (a proxy of cell size) and SYBR Green fluorescence (an indicator of nucleic acid) were used to distinguish viruses from heterotrophic bacteria [29].

All statistical analyses were performed using SPSS software (SPSS 14). A Student *t*-test was carried out to check for significant differences in microbial community (viruses, bacteria, *Synechococcus* spp., HNFs, PNFs, and ciliates) and environmental factors (temperature, salinity, NO_3, and Chl *a*) between the northern and southern parts of the Taiwan Strait, with a confidence level of 95% ($\alpha = 0.05$). Regression analyses were used to explain the relationship between microbial community and environmental factors. This study used redundancy analysis (RDA; rda function in R package *vegan*) to examine the environmental conditions to explain the microbial community composition in the studied parameters and the spatial variation of the microbial community. To identify the main patterns of spatial change of the relationship between microbial community and environmental conditions, we conducted a series of RDAs using (1) all datasets of the Taiwan Strait, (2) northern part, and (3) southern part of the Taiwan Strait.

3. Results

3.1. Environmental Conditions

The spatial variations in hydrographic parameters (temperature, salinity, and NO_3 concentration) in the surface waters of the Taiwan Strait are shown in Figure 1. During the study period, water temperature and salinity were found to have different patterns in different areas, with the northern part of the Taiwan Strait having lower temperatures (15.3–22.8 °C) and salinities (32.2–33.4 psu) (transects A–D) (Figure 1A,B), as they were affected by China's coastal waters. Temperatures and salinities were higher in the southern part (transects E–I) (*t*-test, $p < 0.05$, Figure 2A). NO_3 concentrations were higher (>4 μM) in the northern part compared to the southern part of the study area (Figures 1C and 2B). Chl *a* concentrations were highest (1.6 mg m^{-3}) at the edge of China's coastal waters in the northern part (Figure 1D). In general, the northern and southern parts of the Taiwan Strait had significant differences in average Chl *a* concentrations (*t*-test, $p < 0.05$) (Figure 2B).

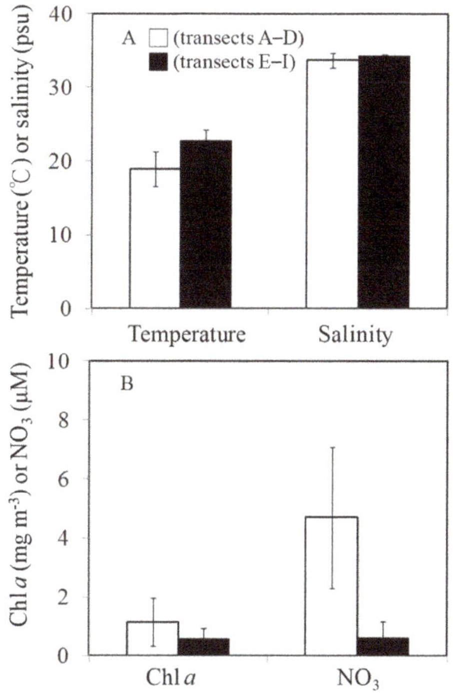

Figure 2. Mean values were determined from the transects A–D (□) and transects E–I (■) for temperature and salinity (**A**) and for Chl *a* and NO_3 (**B**).

3.2. Spatial Changes in Microbial Community

During the study period, the spatial variations in viral abundance ranged from 0.4×10^7 to 2.4×10^7 viruses mL^{-1}, was slightly lower in the northern (transects A–D) part $(1.2 \pm 0.5) \times 10^7$ viruses mL^{-1}) than that in the southern (transects E–I) part of the Taiwan Strait, $(1.5 \pm 0.6) \times 10^7$ viruses mL^{-1}) (Figure 3A). Bacterial abundance varied between 0.4×10^6 and 1.1×10^6 cells mL^{-1} and 0.6×10^6 and 2.9×10^6 cells mL^{-1} in the northern (transects A–D) part and southern (transects E–I) part of the Taiwan Strait, respectively (Figure 3B). Bacterial abundances in the northern part (transects A–D) were significantly lower than those in the southern part (transects E–I) (Figure 4). Based on our field studies, bacterial abundance was probably related to temperature, and temperature explained about 42% of the difference in bacterial abundance (Figure 5A). Furthermore, viral abundance was significantly positively related to bacterial abundance in the northern part ($r = 0.41$, $p < 0.05$) and southern part ($r = 0.82$, $p < 0.05$) of the Taiwan Strait (Figure 5B). Abundance of *Synechococcus* spp. ranged from 0.4×10^4 to 4.2×10^4 cells mL^{-1} and showed higher values in the northern part (transects A–D) (Figures 3C and 4). Nanoflagellate concentrations were higher in the northern part (transects A–D), HNF concentrations were $(0.6–2.8) \times 10^3$ cells mL^{-1}, and PNF concentrations were $(0.3–2.3) \times 10^3$ cells mL^{-1} (Figure 3D,E and Figure 4). When *Synechococcus* spp. and PNFs from all stations were pooled, they were found to be positively correlated with total Chl *a* values (*Synechococcus* spp.: $r = 0.77$, $p < 0.05$; PNF: $r = 0.63$, $p < 0.05$; Figure 6). Mean *Synechococcus* spp. and PNFs represented a significant part of the total phytoplankton biomass. The same pattern was noted for ciliate abundances and the nanoflagellate community, with the southern part of the Taiwan Strait having lower ciliate concentrations (<400 cells L^{-1}) (Figures 3F and 4).

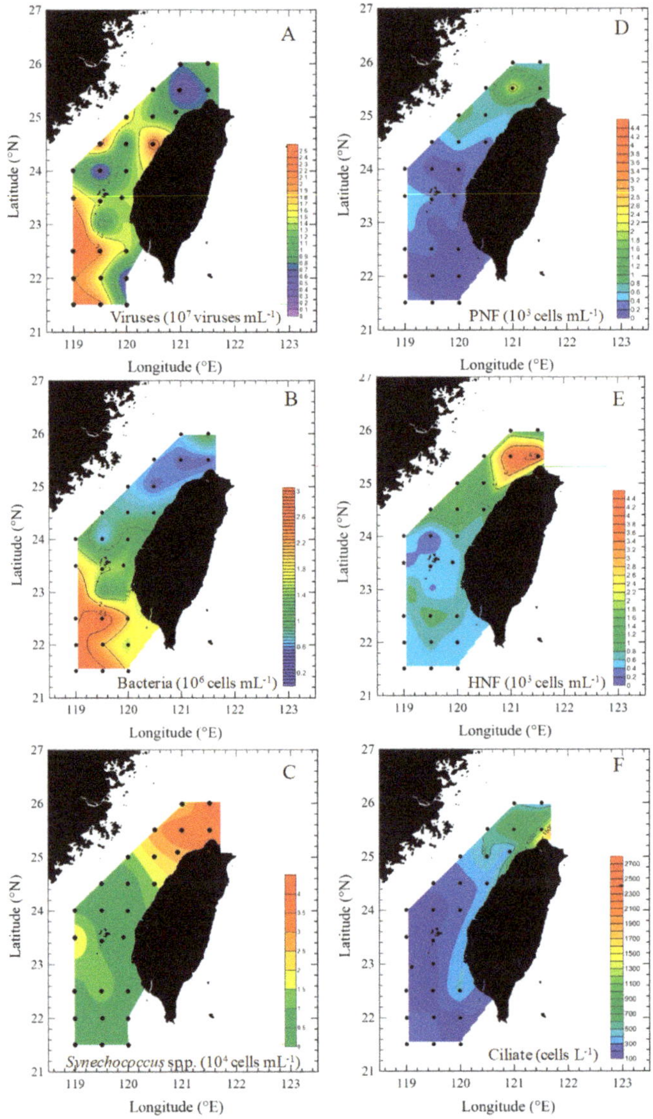

Figure 3. Spatial variations of viral (**A**), bacterial (**B**), *Synechococcus* spp. (**C**), heterotrophic nanoflagellate (**D**), pigmented nanoflagellate (**E**), and ciliate (**F**) abundance during the study period in Taiwan Strait.

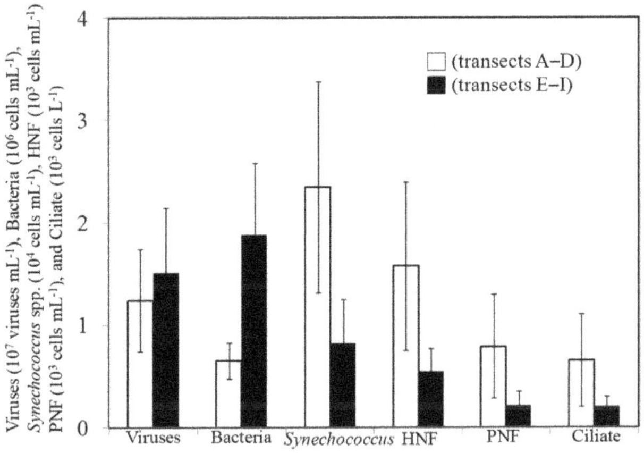

Figure 4. Mean values were determined from the transects A–D (□) and transects E–I (■) for viral, bacterial, *Synechococcus* spp., HNF, PNF, and ciliate abundances.

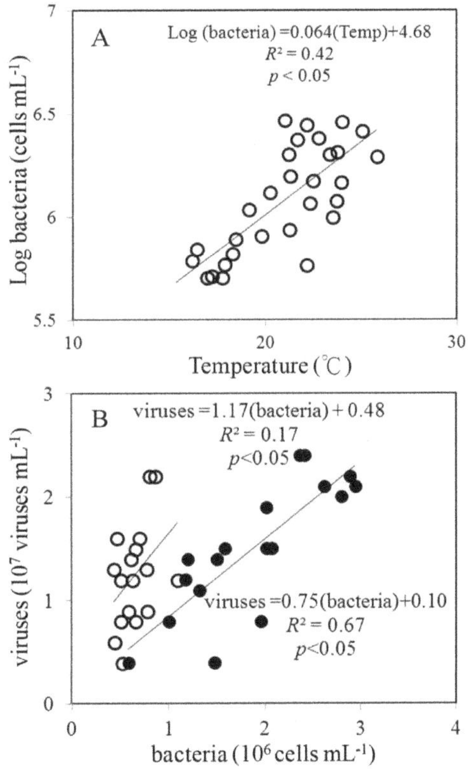

Figure 5. Relationship between temperature and log(bacterial abundance) (**A**) and between viral and bacterial abundance (**B**) during the study period. The solid line is the regression line for all data (○); (□): the data for transects A–D; (■): the data for transects E–I.

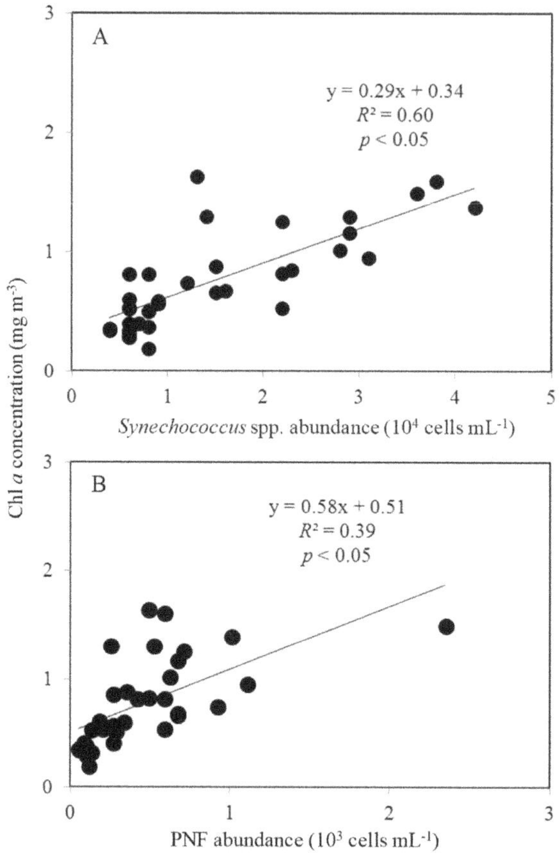

Figure 6. Relationship between *Synechococcus* spp. (**A**), PNF abundance (**B**), and Chl *a* concentration during the study period. The solid line is the regression line for all data.

Based on analysis of all the data, environmental conditions explained 43.7% of variance in the data for microbial community composition in the Taiwan Strait. We observed that HNFs, PNFs, ciliates, and *Synechococcus* spp. showed similar patterns of spatial fluctuations (Figure 7A). On the other hand, temperature and salinity were associated with bacteria (Figure 7A). This result is supported by the correlation analysis between both of them (Figure 5A). Interestingly, a different pattern was observed between the northern and the southern parts of the Taiwan Strait (Figure 7B,C). In the northern part of the Taiwan Strait, we noticed the marked impact of the bacteria characterized by temperature, salinity, and viruses (Figure 7B). However, in the southern part of the Taiwan Strait, the impact parameters appeared in a different way, as bacteria variations were associated with an abundance of viruses and HNFs, highlighting the importance of top-down control as a limiting factor on bacteria (Figure 7C).

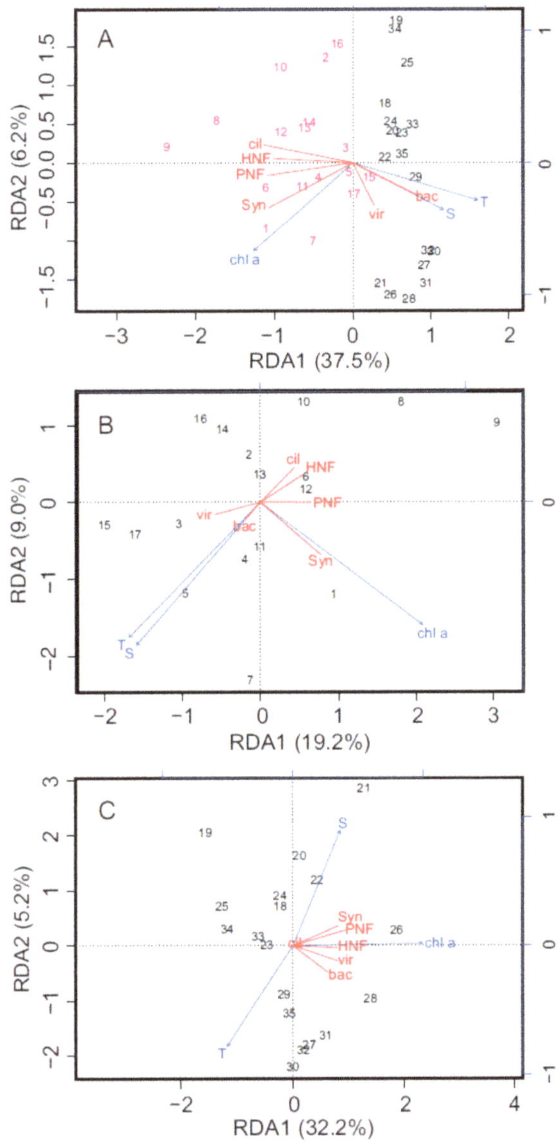

Figure 7. Redundancy analysis (RDA)-based microbial community composition (bac = bacteria, vir = viruses, HNF = heterotrophic nanoflagellate, PNF = pigmented nanoflagellate, cil = ciliate, and Syn = *Synechococcus* spp.), and environmental conditions (chl a = chlorophyll a, T = temperature, S = salinity) as the explanatory variables, for all data (**A**), northern part (**B**), and southern part (**C**) of the Taiwan Strait. In (A), samplings from northern part and southern part of the Taiwan Strait are labeled using magenta and black, respectively. The percentages of the community composition variation explained by the first and second RDA are labeled on the two axes of each RDA plot. The explanatory environmental conditions and microbial compositions are presented as blue arrows and red lines on RDA plots, respectively.

4. Discussion

This study explored spatial variations in microbial communities across the Taiwan Strait during the winter months (January and February) of 2018. We used temperature and salinity in our study period to distinguish two hydrographic regions in the Taiwan Strait. Transects A–D and transects E–I were assumed in this study to be representative of the northern and southern parts of the strait, respectively. To the best of our knowledge, there are no previous studies of aquatic–microbial community dynamics in the Taiwan Strait, and so this study represents the first to investigate aquatic microbial ecology in this region. We found viral and bacterial concentrations to be higher at higher temperatures and in oligotrophic oceans, which have higher salinities in the southern part of the Taiwan Strait. *Synechococcus* spp., nanoflagellates, and ciliates had different spatial variations, with higher abundances of both found in the northern part (transects A–D) of the Taiwan Strait containing cold China coastal waters.

4.1. Spatial Variations in Bacterial and Viral Abundance

In marine environments, substrate availability and water temperature are important factors regulating bacterial production [9,30–33]. Some previous studies have suggested that, over the course of a year, temperature is the dominant factor affecting bacterial growth in colder seasons. Other factors, such as substrate supply, may be more important in warmer seasons [9,34]. In a study of a Delaware estuary, Hock and Kirchman [34] showed that its temperature varied over an annual cycle and that bacterial growth rate was directly related to temperatures below, but not above, 12 °C. In the present study, we found a clear relationship between temperature and bacterial abundance (ANOVA, $p < 0.05$) (Figure 5A), but not between Chl *a* and bacteria concentrations.

Viruses control bacterial mortality and may help to maintain bacterial community diversity [35]. In most aquatic environments, viral abundance was positively correlated with bacteria [27,35,36]. Interestingly, we found a closer linear relationship between viruses and bacteria in oligotrophic oceans in the southern part than those in the northern part of the Taiwan Strait (Figure 5B). One previous study of two lakes in the French Massif Central found that the relationship between viral and bacterial abundances to be weaker in the more productive lake, suggesting that there is an increase in relative abundance of non-bacteriophage viruses, such as cyanophages, in more productive environments [37]. In the present study, *Synechococcus* spp. abundance was significantly higher in the northern part of the study area (transects A–D) (Figure 4). This might explain the weaker linear relationship between the abundance of viruses and bacteria in the northern part of the Taiwan Strait (Figure 5B). Aside from this, differences in these relationships may be related to the burst size and infection rate of host cells [36]. Thus, future studies may want to include viral lysis and viral production when assessing the ecological role of phages and their potential for controlling bacterial dynamics.

4.2. Effects of Bottom-Up and Top-Down Controls on HNFs

According to one study, HNF abundance should be higher in warm environments than in colder ones, owing to consistent differences in the food web structure along the latitudinal gradient [38]. Some models of the predator–prey relationship have predicted that an increase in trophy of the freshwater environment should be reflected in both higher prey and higher predator abundances [39]. For example, a positive relationship has been found between bacterial and HNF abundances in marine systems [12,40–42]. Therefore, we also expected to find a relationship between the abundances of bacteria and HNFs in our study, but after considering all the data we collected, we found no consistent positive relationship between bacterial and HNF abundances (Figure 8), which is a correlation reported previously by many studies [12,41–43]. Šolić et al. [44] also reported a very low association between bacteria and HNF concentrations during the colder part of the year, suggesting that HNFs did not graze as much on bacteria during the colder months in the middle of the Adriatic Sea. In addition, top-down control on HNFs weakens the relationship between the abundances of bacteria and HNFs [15,16]. Therefore, a negative relationship was found between bacteria and HNFs in the northern part of the

Taiwan Strait (r = −0.57) (Figure 8), showing a top-down control on bacteria by HNFs in this study area. The significant and positive relationship between water temperature and bacterial abundance (r = 0.65) suggests a stronger bottom-up control of bacteria in the northern part of the Taiwan Strait (Figure 5A). For another possibility of both bacteria and HNFs showing a negative relationship, the high correlation between HNFs and ciliates (strong top-down control of HNFs) could explain their relationship through the "trophic cascade effect." This predation effect is known as the trophic cascade and is based on predation limitation at several trophic levels [45].

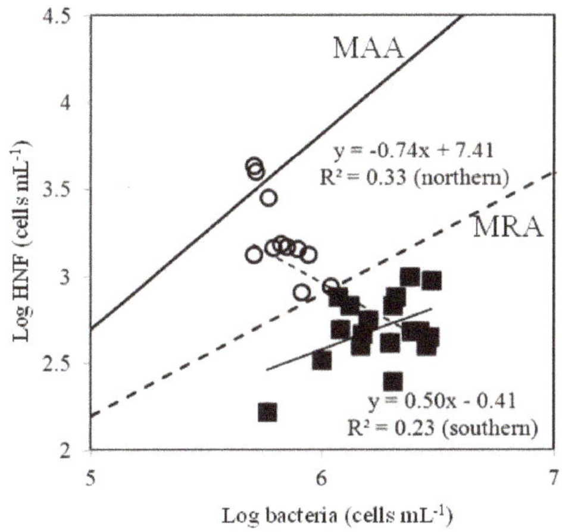

Figure 8. Simultaneous observations of bacterial and HNF abundances plotted following Gasol's model [16]. MAA is the maximum attainable abundance line, and MRA is the mean realized abundance line. (○): the data for transects A–D; (■): the data for transects E–I.

We looked to interpret our data using Gasol's [16] theoretical framework to better understand the HNFs' limitation in our study (Figure 8). According to Gasol's model, points located below the MRA line (mean realized abundance) suggest top-down control of HNFs and points above it suggest low top-down control of HNFs. Points close to the MAA (maximal attainable abundance) line indicate clear bottom-up control of HNFs. The points were all below the MRA line in the southern part of Taiwan Strait in this study, denoting that HNFs were controlled more by predation (top-down) than available resources (bottom-up) (Figure 8). However, we did not observe a higher abundance of ciliates, top-down predators of HNFs, in the southern part of the Taiwan Strait (Figures 3D and 4). However, viral lysis is a major source of bacterial loss [46,47], and its effect on bacterial mortality has been compared with the effect of nanoflagellate grazing in many oceanic systems [47–49].

This study found a closer linear relationship between viruses and bacteria in the southern part than those in the northern part of the Taiwan Strait (Figure 5B), thus suggesting that a significant fraction of bacterial carbon and energy is probably not transferred to higher trophic levels, but instead is cycled in a bacteria–virus–dissolved organic matter (DOM) loop. This "viral loop" [50] might serve as an important control mechanism for bacterial production in the southern part of the Taiwan Strait. Therefore, future studies may want to take viral lysis into account when studying mechanisms controlling bacterial abundance. Aside from this, mixotrophic nanoflagellates (MNFs) and HNFs have been described as important grazers on bacteria; in particular, the relevance of MNFs as grazers in oligotrophic environments was recognized [8]. Furthermore, a group of grazers (MNFs) may be related to bacteria, but not to HNFs, which probably affects the bacteria–HNF relationship.

We applied our field data of bacteria and HNF abundances in the northern part of the Taiwan Strait to the model by Gasol [16] (Figure 8). Because most points lay between the MRA and MAA lines in this part of the strait in this study, top-down control on HNFs should be low and bottom-up control should be high (Figure 8). This part of the Taiwan Strait is relatively high in nutrients. Thus, our results were not in agreement with Gasol regarding dominant bottom-up control of HNFs in oligotrophic sites [16]. At the stations in the northern part of the strait, bacterial abundance was relatively low, because low temperature reduced bacterial growth. In this situation, HNFs could not satisfy their carbon demand through only grazing on bacteria in the northern part of the strait.

The increase in HNF abundance in the northern part of Taiwan Strait may have to do with the availability of other types of prey, such as *Synechococcus* spp. or picoeukaryotes, as well as low top-down control by predators such as ciliates. Some studies have found variations in the abundance of ciliates to be important in explaining the variations in abundance of nanoflagellates. Weisse [51] and Nakano et al. [52] reported a negative relationship between abundances of the two. According to earlier studies, predation control on HNFs should increase with productivity [12,16]. However, in the current study, we found a positive relationship between ciliates and nanoflagellates (ANOVA, $p < 0.05$). As for the ciliate community, ciliates of <30 μm in equivalent spherical diameter (ESD), including *Strombidium* spp., *Strobilidium* spp., and *Tontonia* spp., were the most abundant group (65–90%) of the ciliate community in the northern part of the strait in this study. These small ciliates probably control the production of picoplankton [53]. This positive relationship between ciliates and nanoflagellates may suggest that the same factors control the abundance of both communities, since they basically have the same resource requirements (picophytoplankton) and potential predators (zooplankton) [21]. Thus, the high abundance of HNFs obtained in relation to the low abundance of bacteria we found in the northern part of the strait in this study was probably due to the availability of other prey sources. HNFs are known to have other food resources (e.g., picophytoplankton) besides bacteria [54]. However, this does not change the fact that we found that HNFs were primarily resource limited in the northern part of the study area. The spatial variations in *Synechococcus* spp. abundance in the present study showed higher values in the northern part (transects A–D) (Figure 4). Thus, we believe that the major flux in energy in the northern part of the strait during winter was nanofalelltes grazing on picophytoplankton.

5. Conclusions

We found higher abundances of bacteria in waters with warmer temperatures and lower salinities, while nanoflagellate and ciliate abundances were found to be higher in the northern part of the Taiwan Strait due to cold China coastal waters. Our findings characterize two different ecosystems in the Taiwan Strait. We suspect that a "viral loop" plays a major role in controlling bacterial production in the southern part of the strait and that picophytoplankton plays a major role in the flux of energy in the northern part of the strait during winter, though further studies are needed to investigate these possibilities.

Author Contributions: Conceptualization: G.-C.G. and A.-Y.T.; methodology: A.-Y.T.; validation: A.-Y.T.; formal analysis: G.-C.G., C.-h.H., P.-C.H., and A.-Y.T.; investigation: H.-M.Y. and Y.-K.C.; resources: G.-C.G. and A.-Y.T.; data curation: G.-C.G. and A.-Y.T.; writing–original draft preparation: G.-C.G. and A.-Y.T.; writing–review and editing: G.-C.G. and A.-Y.T.; supervision: G.-C.G.; funding acquisition: G.-C.G. and A.-Y.T.

Funding: This study was supported by a grant (NSC 106-2611-M-019-013) from the Ministry of Science and Technology, Republic of China.

Acknowledgments: We appreciate the language editing and helpful comments from James Steed on this manuscript.

Conflicts of Interest: The authors declare no conflict of interest.

References

1. Pomeroy, L.R. The ocean's food web: A changing paradigm. *Bioscience* **1974**, *24*, 499–504. [CrossRef]
2. Christaki, U.; Giannakourou, A.; Van Wambeke, F.; Gregori, G. Nanoflagellate predation on auto-and heterotrophic picoplankton in the oligotrophic Mediterranean Sea. *J. Plankton Res.* **2001**, *23*, 1297–1310. [CrossRef]
3. Tsai, A.Y.; Chiang, K.P.; Chang, J.; Gong, G.C. Seasonal diel variations of picoplankton and nanoplankton in a subtropical Western Pacific coastal ecosystem. *Limnol. Oceanogr.* **2005**, *50*, 1221–1231. [CrossRef]

4. Unrein, F.; Massana, R.; Alonso-Saez, L.; Gasol, J.M. Significant year-round effect of small mixotrophic flagellates on bacterioplankton in an oligotrophic coastal system. *Limnol. Oceanogr.* **2007**, *52*, 456–469. [CrossRef]
5. Tsai, A.Y.; Gong, G.C.; Hung, J. Seasonal variations of virus- and nanoflagellate-mediated mortality of heterotrophic bacteria in the coastal ecosystem of subtropical western Pacific. *Biogeosciences* **2013**, *10*, 3055–3065. [CrossRef]
6. Schultz, G.E., Jr.; White, E.D., III; Ducklow, H.W. Bacterioplankton dynamics in the York River estuary: Primary influence of temperature and freshwater inputs. *Aquat. Microb. Ecol.* **2003**, *30*, 135–148. [CrossRef]
7. Iriarte, A.; Sarobe, A.; Orive, E. Seasonal variability in bacterial abundance, production and protistan bacterivory in the lower Urdaibai estuary, Bay of Biscay. *Aquat. Microb. Ecol.* **2008**, *52*, 273–282. [CrossRef]
8. Tsai, A.Y.; Gong, G.C.; Sanders, R.W.; Chen, W.H.; Chao, C.F.; Chiang, K.P. Importance of bacterivory by pigmented and heterotrophic nanoflagellates during the warm season in a subtropical western Pacific coastal ecosystem. *Aquat. Microb. Ecol.* **2011**, *63*, 9–18. [CrossRef]
9. Li, W.K.W. Annual average abundance of heterotrophic bacteria and *Synechococcus* in surface ocean waters. *Limnol. Oceanogr.* **1998**, *43*, 1746–1753. [CrossRef]
10. Ochs, C.A.; Cole, J.J.; Liken, G.E. Population dynamics of bacterioplankton in an oligotrophic lake. *J. Plankton Res.* **1995**, *17*, 365–391. [CrossRef]
11. Shiah, F.K.; Ducklow, H.W. Temperature regulation of heterotrophic bacterioplankton abundance, production, and specific growth rate in Chesapeake Bay. *Limnol. Oceanogr.* **1994**, *39*, 1243–1258. [CrossRef]
12. Sanders, R.W.; Caron, D.A.; Berninger, U.G. Relationships between bacteria and heterotrophic nanoplankton in marine and fresh waters- an interecosystem comparison. *Mar. Ecol. Peog. Ser.* **1992**, *86*, 1–14. [CrossRef]
13. Sherr, E.B.; Sherr, B.F. Significance of predation by protists in aquatic microbial food webs. *Antonie Van Leeuwenhoek* **2002**, *81*, 293–308. [CrossRef] [PubMed]
14. Gasol, J.M.; Simons, A.M.; Kalff, J. Patterns of top-down versus bottom-up regulation of heterotrophic nanoflagellates in temperate lakes. *J. Plankton Res.* **1995**, *17*, 1879–1903. [CrossRef]
15. Gasol, J.M.; Vaque, D. Lack of coupling between heterotrophic nanoflagellates and bacteria: A general phenomenon across aquatic systems? *Limnol. Oceanogr.* **1993**, *38*, 657–665. [CrossRef]
16. Gasol, J.M. A framework for the assessment of top-down vs. bottom-up control of heterotrophic nanoflagellate abundance. *Mar. Ecol. Prog. Ser.* **1994**, *113*, 291–300. [CrossRef]
17. Kisand, V.; Zingel, P. Dominance of ciliate grazing on bacteria during spring in a shallow eutrophic lake. *Aquat. Microb. Ecol.* **2000**, *22*, 135–142. [CrossRef]
18. Berglund, J.; Samuelsson, K.; Kull, T.; Muren, U.; Andersson, A. Relative strength of resource and predation limitation of heterotrophic nanoflagellates in a low-productive sea area. *J. Plankton Res.* **2005**, *27*, 923–935. [CrossRef]
19. Samuelsson, K.; Berglund, J.; Andersson, A. Factors structuring the heterotrophic flagellate and ciliate community along a brackish water primary production gradient. *J. Plankton Res.* **2006**, *28*, 345–359. [CrossRef]
20. Carrias, J.F.; Amblard, C.; Quiblier-Lloberas, C.; Bourdier, G. Seasonal dynamics of free and attached heterotrophic nanoflagellates in an oligotrophic lake. *Freshw. Biol.* **1998**, *39*, 91–101. [CrossRef]
21. Auer, B.; Arndt, H. Taxonomic composition and biomass of heterotrophic flagellates in relation to lake trophy and season. *Freshw. Biol.* **2001**, *46*, 959–972. [CrossRef]
22. Kosolapova, N.G.; Kosolapov, D.B. The diversity and distribution of heterotrophic nanoflagellates in the eutrophic Lake Nero. *Inland Water Biol.* **2009**, *2*, 42–49. [CrossRef]
23. Liang, W.D.; Tang, T.Y.; Yang, Y.J.; Ko, M.T.; Chung, W.S. Upper-ocean currents around Taiwan. *Deep Sea Res.* **2003**, *50*, 1085–1105. [CrossRef]
24. Jan, S.; Wang, J.; Chern, C.S.; Chao, S.Y. Seasonal variation of the circulation in the Taiwan Strait. *J. Mar. Syst.* **2002**, *35*, 249–268. [CrossRef]
25. Gong, G.C.; Shiah, F.-K.; Liu, K.-K.; Wen, Y.-H.; Liang, M.-H. Spatial and temporal variation of chlorophyll a, primary productivity and chemical hydrography in the southern East China Sea. *Cont. Shelf Res.* **2000**, *20*, 411–436. [CrossRef]
26. Stoecker, D.K.; Taniguchi, A.; Michaels, A.E. Abundance of autotrophic, mixotrophic, and heterotrophic planktonic ciliates in shelf and slope waters. *Mar. Ecol. Prog. Ser.* **1989**, *50*, 241–254. [CrossRef]
27. Tsai, A.Y.; Gong, G.C.; Liu, H. Seasonal variations in virioplankton and picoplankton in semi-enclosed and open coastal waters. *Terr. Atoms. Ocean. Sci.* **2018**, *29*, 465–472. [CrossRef]

28. Porter, K.G.; Feig, Y.S. The use of DAPI for identifying and counting aquatic microflora. *Limnol. Oceanogr.* **1980**, *25*, 943–948. [CrossRef]
29. Brusaard, C.P.D. Optimization of procedures for counting viruses by flow cytometry. *Appl. Environ. Microbial.* **2004**, *70*, 1506–1513. [CrossRef]
30. White, P.A.; Kalff, J.; Rasmussen, J.B.; Gasol, J.M. The effect of temperature and algal biomass on bacterial production and specific growth rate in freshwater and marine habitats. *Microb. Ecol.* **1991**, *21*, 99–118. [CrossRef]
31. Coveney, M.F.; Wetzel, R.G. Effects of nutrients on specific growth rate of bacterioplankton in oligotrophic lake water cultures. *Appl. Environ. Microbiol.* **1992**, *58*, 150–156. [PubMed]
32. Sala, M.M.; Peters, F.; Gasol, J.M.; Pedros-Alio, C.; Marrase, C.; Vaque, D. Seasonal and spatial variations in the nutrient limitation of bacterioplankton growth in the northwestern Mediterranean. *Aquat. Microb. Ecol.* **2002**, *27*, 47–56. [CrossRef]
33. Rejas, D.; Muylaert, K.; Meester, L.D. Nutrient limitation of bacteria and sources of nutrients supporting nutrient-limited bacterial growth in an Amazonina floodplain lake. *Aquat. Microb. Ecol.* **2005**, *39*, 57–67. [CrossRef]
34. Hock, M.P.; Kirchman, D.L. Seasonal and inter-annual variability in bacterial production and biomass in a temperate estuary. *Mar. Ecol. Prog. Ser.* **1993**, *98*, 283–295. [CrossRef]
35. Peduzzi, P.; Schiemer, F. Bacteria and viruses in the water column of tropical freshwater reservoirs. *Environ. Microbiol.* **2004**, *6*, 707–715. [CrossRef] [PubMed]
36. Wommack, K.E.; Colwell, R.R. Virioplankton: Viruses in aquatic ecosystems. *Microbiol. Mol. Biol. Rev.* **2000**, *64*, 69–114. [CrossRef]
37. Bettarel, Y.; Sime-Ngando, T.; Amblard, C.; Dolan, J. Viral activity in two contrasting lake ecosystems. *Appl. Environ. Microbiol.* **2004**, *70*, 2941–2951. [CrossRef]
38. Sarmento, H. New paradigms in tropical limnology: The importance of the microbial food web. *Hydrobiologia* **2012**, *686*, 1–14. [CrossRef]
39. Arditi, R.; Ginzburg, L.R.; Akcakaya, H.R. Variation in plankton densities among lakes: A case for ratio-dependent predation models. *Am. Nat.* **1991**, *138*, 1287–1296. [CrossRef]
40. Sherr, B.F.; Sherr, E.B.; Newell, S.Y. Abundance and productivity of heterotrophic nanoplankton in Georgia coastal waters. *J. Plankton Res.* **1984**, *6*, 195–202. [CrossRef]
41. Wright, R.T.C.; Wright, R.T.; Ledo, M.E. Dynamics of planktonic bacteria and heterotrophic flagellates in Parker Estuary, northern Massachusetts. *Cont. Shelf Res.* **1987**, *7*, 1383–1387. [CrossRef]
42. Wikner, J.; Hagstrom, A. Evidence of a tightly coupled nanoplankton predator-prey link regulating the bacterivores in the marine environment. *Mar. Ecol. Prog. Ser.* **1988**, *50*, 137–145. [CrossRef]
43. Berninger, U.G.; Finlay, B.J.; Kuuppo-Leinikki, P. Protozoan control of bacterial abundances in freshwater. *Limnol. Oceanogr.* **1991**, *36*, 139–147. [CrossRef]
44. Solic, M.; Krstulovic, N.; Vilibic, I.; Bojanic, N.; Kuspilic, G.; Sestanovic, S.; Santic, D.; Ordulj, M. Variability in the bottom-up and top-down controls of bacteria on trophic and temporal scales in the middle Adriatic Sea. *Aquat. Microb. Ecol.* **2009**, *58*, 15–29. [CrossRef]
45. Paine, R.T. Food Webs: Linkage, interaction strength and community infrastructure. *J. Anim. Ecol.* **1980**, *49*, 667–685. [CrossRef]
46. Suttle, C.A. The significance of viruses to mortality in aquatic microbial communities. *Microb. Ecol.* **1994**, *28*, 237–243. [CrossRef] [PubMed]
47. Fuhrman, J.A.; Noble, R.T. Viruses and protists cause similar bacterial mortality in coastal seawater. *Limnol. Oceanogr.* **1995**, *40*, 1236–1242. [CrossRef]
48. Steward, F.C.; Smith, D.C.; Azam, F. Abundance and production of bacteria and viruses in the Bering and Chukchi Sea. *Mar. Ecol. Prog. Ser.* **1996**, *131*, 287–300. [CrossRef]
49. Taira, Y.; Uchimiya, M.; Kudo, I. Simultaneous estimation of viral lysis and protozoan grazing on bacterial mortality using a modified virus-dilution method. *Mar. Ecol. Prog. Ser.* **2009**, *379*, 23–32. [CrossRef]
50. Bratbak, G.; Heldal, M.; Thingstad, T.F.; Riemann, B.; Haslund, O.H. Incorporation of viruses into the budget of microbial C-transfer. A first approach. *Mar. Ecol. Prog. Ser.* **1992**, *83*, 273–280. [CrossRef]
51. Weisse, T. The annual cycle of heterotrophic freshwater nanoflagellates: Role of bottom-up versus top-down control. *J. Plankton Res.* **1991**, *13*, 167–185. [CrossRef]

52. Nakano, S.P.M.; Manage, M.; Nishibe, Y.; Kawabata, Z. Trophic linkage among heterotrophic nanoflagellates, ciliates and metazoan zooplankton in a hypereutrophic pond. *Aquat. Microb. Ecol.* **2001**, *25*, 259–270. [CrossRef]
53. Suzuki, T.; Miyabe, C. Ecological balance between ciliate plankton and its prey candidates, pico- and nanoplankton, in the East China Sea. *Hydrobiology* **2007**, *586*, 403–410. [CrossRef]
54. Kuosa, H. Picoplanktonic algae in the northern Baltic Sea-seasonal dynamics and flagellate grazing. *Mar. Ecol. Prog. Ser.* **1991**, *73*, 269–276. [CrossRef]

© 2019 by the authors. Licensee MDPI, Basel, Switzerland. This article is an open access article distributed under the terms and conditions of the Creative Commons Attribution (CC BY) license (http://creativecommons.org/licenses/by/4.0/).

diversity

MDPI

Article

Changes in Heterotrophic Picoplankton Community Structure after Induction of a Phytoplankton Bloom under Different Light Regimes

Hera Karayanni [1,*], Konstantinos A. Kormas [2], Maria Moustaka-Gouni [3] and Ulrich Sommer [4]

1 Department of Biological Applications and Technology, University of Ioannina, 45110 Ioannina, Greece
2 Department of Ichthyology and Aquatic Environment, School of Agricultural Sciences, University of Thessaly, 38446 Volos, Greece; kkormas@uth.gr
3 School of Biology, Department of Botany, Aristotle University of Thessaloniki, 54124 Thessaloniki, Greece; mmustaka@bio.auth.gr
4 Helmholtz Center for Ocean Research (GEOMAR), Dusternbrooker Weg 20, 24105 Kiel, Germany; usommer@geomar.de
* Correspondence: hkaray@uoi.gr; Tel.: +30-2651-007341

Received: 20 September 2019; Accepted: 11 October 2019; Published: 15 October 2019

Abstract: Bacterial and archaeal diversity and succession were studied during a mesocosm experiment that investigated whether changing light regimes could affect the onset of phytoplankton blooms. For this, 454-pyrosequencing of the bacterial V1–V3 and archaeal V3–V9 16S rRNA regions was performed in samples collected from four mesocosms receiving different light irradiances at the beginning and the end of the experiment and during phytoplankton growth. In total, 46 bacterial operational taxonomic units (OTUs) with ≥1% relative abundance occurred (22–34 OTUs per mesocosm). OTUs were affiliated mainly with Rhodobacteraceae, Flavobacteriaceae and Alteromonadaceae. The four mesocosms shared 11 abundant OTUs. Dominance increased at the beginning of phytoplankton growth in all treatments and decreased thereafter. Maximum dominance was found in the mesocosms with high irradiances. Overall, specific bacterial OTUs had different responses in terms of relative abundance under in situ and high light intensities, and an early phytoplankton bloom resulted in different bacterial community structures both at high (family) and low (OTU) taxonomic levels. Thus, bacterial community structure and succession are affected by light regime, both directly and indirectly, which may have implications for an ecosystem's response to environmental changes.

Keywords: bacterial and archaea diversity; light; climate change; phytoplankton blooms; 454-pyrosequencing

1. Introduction

It is well known today that one of the most important pools of dissolved organic matter (DOM) in the euphotic layer is of phytoplankton origin. Heterotrophic bacteria are considered to be highly responsive to environmental perturbations that affect phytoplankton growth and, subsequently, DOM release [1] due to their small size, life cycle, dispersal potential and metabolic capabilities [2,3]. However, interactions are amphidromous—the composition of the DOM released by phytoplankton may affect bacterial growth and community structure, but bacteria will also affect its amount and composition [4]. The influence of phytoplankton on bacteria may be also exerted through other mechanisms like competition for nutrients and release of antibiotic compounds [5], and effects can be taxon specific (i.e., different phytoplankton taxa may affect positively or negatively specific bacteria [6,7]). Thus, in a changing ocean it is important to understand the complex interactions between phytoplankton and bacteria that affect the fate of organic carbon in the biosphere.

Bacterial community structure and succession during bloom events has been previously studied both in microcosm/mesocosm experiments [8] and in situ [9]. These studies have indicated that certain bacterial groups such as Rhodobacterales, SAR11, Alteromonadales and Bacteroidetes [8–10] are often associated with algal blooms or the degradation of the derived organic matter. The use of microcosms and mesocosms is maybe the most suitable experimental methodology available to date when intending to study the impact of environmental factors and climate change on plankton communities [11–13]. Experimental approaches seem to be more informative in terms of species-specific responses to environmental perturbations for the majority of the studied marine [14] and freshwater [5] habitats as compared with field studies. Whether it is about pulse-like events such as extreme meteorological activity, or press-like disturbances such ongoing climatic changes (e.g., CO_2 increase and ocean acidification), prokaryotic communities are expected to respond, either directly or indirectly, and show some degree of resilience both at the structural and functional levels [14,15] in facilitating ecosystem recovery [16].

Light as a driver of ecosystem dynamics has been explored in a number of studies, but has generally received less attention compared to other environmental factors like temperature and nutrients. It has been shown, however, that light may have direct or indirect effects at cellular to community levels [17–20]. For instance, stimulation of bacterial clades containing aerobic anoxygenic phototrophs (*Rhodobacteraceae* and NOR5) and inhibition of SAR11 [21–23] have been shown. Although this is considered to be a general trend, members of the aforementioned clades may exhibit differential responses to light manipulation [24]. Similarly, exposure of marine isolates to sea surface solar irradiance has indicated a large interspecific variability of resistance [25]. Indirect effects of light on prokaryotes are mainly associated with their role as major consumers of dissolved organic matter (DOM) [26]. The exudates released by phytoplankton can differ based on light availability [27], while photoalterations to DOM may change its bioavailability [28].

In a mesocosm experiment carried out with natural plankton assemblages from the Western Baltic Sea, it was shown that a phytoplankton bloom can be induced by intermittent high-light periods as early as in January [19]. Additionally, these authors found pronounced changes in the taxonomic response of phytoplankton and copepods nauplii. However, when it comes to heterotrophic prokaryotic members of the food web, there is a knowledge gap. Thus, in this paper, using the same mesocosms as in Sommer et al. [19], we investigated whether changes in the onset of the phytoplankton bloom and the succession of phytoplankton may also be followed by changes in heterotrophic prokaryoplankton community composition due to alterations in light regime. For this, we investigated prokaryoplankton abundance (epifluorescence microscopy) and diversity (16S rDNA 454 pyrosequencing) during the above mentioned 24-day indoor mesocosm experiment simulating climate-driven changes in light intensity. In particular, bacterial community dynamics were studied in two control mesocosms simulating the in situ conditions and light intensity that occur in January and February in coastal North Central Europe. Climate-driven changes (i.e., changes in cloudiness) were simulated by a ~2 fold increase in solar radiation in two additional mesocosms. A phytoplankton bloom occurred in all mesocosms except in the January control, and diatom contribution to total biomass increased with the light dose, a response driven by the predominance of *Skeletonema costatum*.

Although it is difficult to disentangle all the factors that may influence the structure of the bacterial community in the mesocosms, we expect that different light regimes will affect bacterial/archaeal communities directly or indirectly. A previous study dealing with the role of temperature showed a faster bloom development at warmer temperatures as well as a shift of bacterial community structure (BCS) towards the appearance of additional taxa [29]. Interestingly, a simultaneous study of temperature and light effects on bacterial communities' composition indicated that light effects are more likely to be direct, while those of temperature are mediated by phytoplankton [20]. We expect that although prokaryoplankton communities are initially similar in all mesocosms, they will subsequently diverge, and specific taxa are likely to emerge following the quantitative and qualitative changes of phytoplankton assemblages or due to direct effects of light.

2. Materials and Methods

The experimental set-up has been described previously in full [19]. In brief, plankton succession was induced in mesocosms of ~1.5 m^3 containing whole water natural phytoplankton and mesozooplankton communities from the Kiel Bight (Baltic Sea) at in situ temperature conditions (4 °C). The mesocosms were subjected to simulated high-light episodes: five days after the experiment started light intensity increased two-fold over controls for 10 days. Controls simulated January and February under in situ conditions and light intensity. They received a light regime based on natural seasonal irradiance dimmed to 43% E_0 of cloudless days. The initial daily light dose was ~2 and 3 mol photons PAR m^{-2} d^{-1} corresponding to mid-January and mid-February values.

Samples for the current study were retrieved from one mesocosm from each treatment, designated as JC, JH, FC, FH and corresponding to January (J) or February (F), control (C) or high-light (H) conditions. The collection of replicated samples was not feasible in order to minimize the removal of excess water volume for all the samples of the project. The dynamics of bacterioplankton were monitored for a total of 24 days. The whole experiment took place at the indoor mesocosm premises of GEOMAR, Kiel, Germany, in January/February 2010.

2.1. *Sampling and Cell Counts*

Heterotrophic prokaryotic cell counts were determined at two- to three-day intervals throughout the whole incubation experiment. From each mesocosm, 10–15 mL of water was fixed at a final concentration of 2% formaldehyde and kept at 4 °C in the dark for no more than two days. A subsample of 2–10 mL was filtered on black pore-sized PTFE filters of 0.2 μm (Millipore, MA, USA) and stained with DAPI (4′,6-diamidino-2-phenylindole). After mounting the filters on glass slides, the cells were counted with an Axiostar (Zeiss) epifluorescence microscope at 1000× magnification.

Water samples were collected from each mesocosm at the beginning of the experiment (day 1), at the beginning (day 11) and the end (day 15) of the exponential growth phase of phytoplankton and a few days (day 19) before the end of the incubation period. At each sampling point, a total of 500–1000 mL was collected in sterile glass bottles, screened through a 20-μm nylon mesh net (Millipore, MA, USA) then filtered on a 47-mm diameter 0.2 μm pore-sized polycarbonate filter (Millipore, MA, USA), under low vacuum pressure (≤150 mm Hg). The filters were stored at −80 °C in sterile cryovials until DNA extraction.

2.2. *Molecular Analysis*

DNA was extracted using the PowerSoil DNA isolation kit (MoBio Laboratories, CA, USA) according to the manufacturer's protocol after slicing the filters under sterile conditions. DNA concentrations ranged from 2.9–113.7 ng μL^{-1}, based on NanoDrop measurements (A260/A280 ratios 1.60–2.01). DNA was amplified with the primers 27F (5′-AGRGTTTGATCMTGGCTCAG-3′) and 519R (5′-GTNTTACNGCGGCKGCTG-3′) targeting the V1–V3 region of the bacterial 16S rRNA gene and with primers 349f (5′-GYGCASCAGKCGMGAAW-3′) and 915r (5′-GTGCTCCCCCGCCAATTCCT-3′) targeting the V3–V9 region of the archaeal 16S rRNA gene (MRDNA, TX, USA). Amplicon pyrosequencing (bTEFAP) was performed as described in [30]. In brief, a one-step 30-cycle PCR was applied using HotStarTaq Plus Master Mix Kit (Qiagen, Valencia, CA). PCR conditions included: 94 °C for 3 min, then followed by 28 cycles of 94 °C for 30 s; 53 °C for 40 s; 72 °C for 1 min; and a final elongation step at 72 °C for 5 min. Following PCR, all amplicon products from different samples were mixed in equal concentrations and purified using Agencourt Ampure beads (Agencourt Bioscience Corporation, MA, USA). Samples were sequenced utilizing Roche 454 FLX titanium instruments and reagents following the manufacturer's guidelines.

2.3. Data Processing and Analysis

Sequencing data were analyzed using MOTHUR 1.28.0 software [31]. In brief, flowgrams from the individual samples were separated according to their TAG, then denoised using PyroNoise software [32]. After removing primer sequences, TAG and key fragments, only sequences ≥200 bp long with homopolymers shorter than 8 bp were considered for further analysis. Chimeric sequences were recognized and removed using the UCHIME software [33]. A 97% similarity cut-off limit was used for clustering the remaining sequences into operational taxonomic units (OTU). Singletons (i.e., sequences that occurred only once in the whole dataset) were removed from downstream analyses, as they were considered most likely sequencing artifacts [34–36]. The data were standardized based on the smallest sample with permutation [37] in order to minimize differences in sequencing depths between samples. Taxonomic affiliation was assigned according to the SILVA 123 SSU RNA database [38]. Sequences from this study have been submitted to the GenBank BioProject (PRJNA238858).

Diversity indices (Dominance and Shannon H) in the four mesocosms during the course of the experiment were calculated using PAST software v3.25 [39]. Dominance calculated as the 1-Simpson diversity index ranges between 0 (when all OTUs are equally presented) and 1 (dominance of one OTU). The Shannon index accounts for both abundance and evenness. Thus, high Shannon index values characterize communities with many taxa equally distributed. The relative abundance of OTUs in all mesocosms were correlated with the biomass of the dominant phytoplankton groups retrieved from [19]. Correlation analysis based on the Spearman coefficient was also performed using the PAST v3.25 software.

Relationships of the 10 OTUs that occurred in all samples (i.e., for all four mesocosms and at all sampling points) were visualized by constructing association networks based on the Spearman correlation coefficient of their log-transformed relative abundances. Only the statistically significant correlations ($p < 0.02$) were used for the network matrix. Network visualization was performed with Cytoscape 2.8.3 software [40]. Positive correlations between OTUs are indicative of non-competitive relationships as a response to the different mesocosm treatments.

3. Results

3.1. Heterotrophic Prokaryotescell Abundance

Total bacterial and archaeal abundance (Figure 1) on day 1 ranged from 0.57×10^5 to 3.49×10^5 cells mL^{-1}. From that point and until day 11 of the experiment, the prokaryotic cell counts remained fairly constant (0.30×10^5 to 1.26×10^5 cells mL^{-1}). Maximum cell abundance ranged between 3.82×10^5 and 6.90×10^5 cells mL^{-1} and occurred on day 19 for all but the JC mesocosm, where peak abundance was monitored on day 24. In both cases peak abundance occurred during the post-bloom period.

3.2. Diversity and Dynamics of Bacteria and Archaea

A total of 120,003 reads met the quality control standards and corresponded to 1,432 unique bacterial OTUs (97% similarity). In the whole dataset, 629 of the unique OTUs appeared only once (i.e., singletons), and were excluded due to the possibility of sequencing errors [34,36], leaving 803 bacterial OTUs for further analysis. Rarefaction curves (Figure S1) varied among samples and only in a few cases did they start to plateau. Archaeal amplifiable DNA and a high total number of reads (i.e., >2000 per sample, after quality processing) were feasible only for one sample from low light and three samples from high light intensities. Thus archaea were excluded from further analysis. The number of unique bacterial OTUs in each mesocosm (Figure 1) decreased from day 1 (136–368) to day 19 (26–137), but followed different patterns in each mesocosm. Maximum dominance (D) occurred at day 11 in all mesoscosms (Figure 2). Diversity was high at day 1 in all treatments, decreased afterwards and increased again to the initial levels towards the end of the experiment, except for FH (Figure 2).

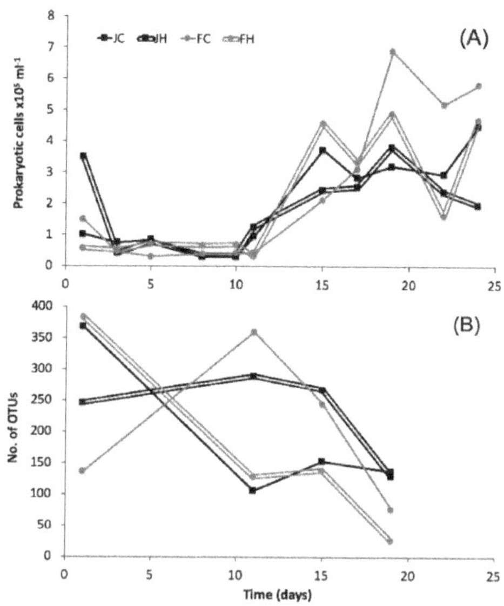

Figure 1. (**A**) Abundance of prokaryotic cells and (**B**) number of unique bacterial operational taxonomic units (OTU) per sample in the studied mesocosms. J: January, F: February, C: control (baseline light intensity), H: high light intensity.

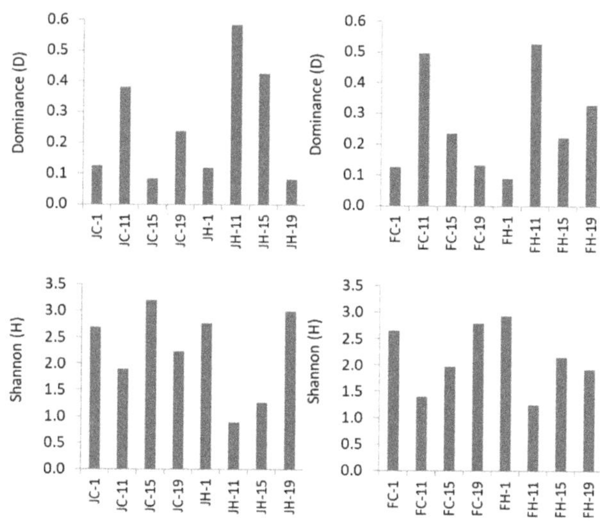

Figure 2. Dominance (D, top) and diversity (Shannon H, bottom) indices of bacterial OTUs in the four mesocosms during the course of the experiment (days 1 to 19). J: January, F: February, C: control, H: high light intensity.

In each mesocosm, 22–34 bacterial OTUs comprised ≥1% of the total abundance. In total, 46 OTUs with ≥1% relative abundance occurred from all mesocosms considered. Analysis was focused on these OTUs, considered as the most abundant ones (Table S1). Eleven of these 46 OTUs appeared in all mesocosms (Figure 3) at various time points. In most of the mesocosms and at most sampling points, the majority of the found OTUs were affiliated to Rhodobacteraceae (Figure S4), with Flavobacteriaceae and Alteromonadaceae being the second-most abundant. The rest of the OTUs belonged to the Verrucomicrobiaceae, Microbacteraceae, Actinomycetales and a few unaffiliated taxa.

Of the 11 OTUs that appeared in all mesocosms, OTU2 and OTU1 were present in all mesocosms at all sampling points and exhibited distinguished temporal patterns (Figure 3). These OTUs belonged to the Rhodobacteraceae family and were related to environmental sequences from coastal waters (Figure S2). The abundance of OTU2 peaked at day 11 in all mesocosms and then declined, with the exception of mesocosm JC, where an increase took place at day 19. On the contrary, the abundance of OTU1 in all mesocosms decreased slightly from the beginning of the experiment and started to increase at the end of the experiment, with the exception of mesocosm JH. The rest of the nine OTUs showed practically no changes in their relative abundance. Another 15 and 5 OTUs were common in three and two mesocosms, respectively, while the last 15 OTUs appeared only in one of the mesocosms (Table S1 and Figure 3).

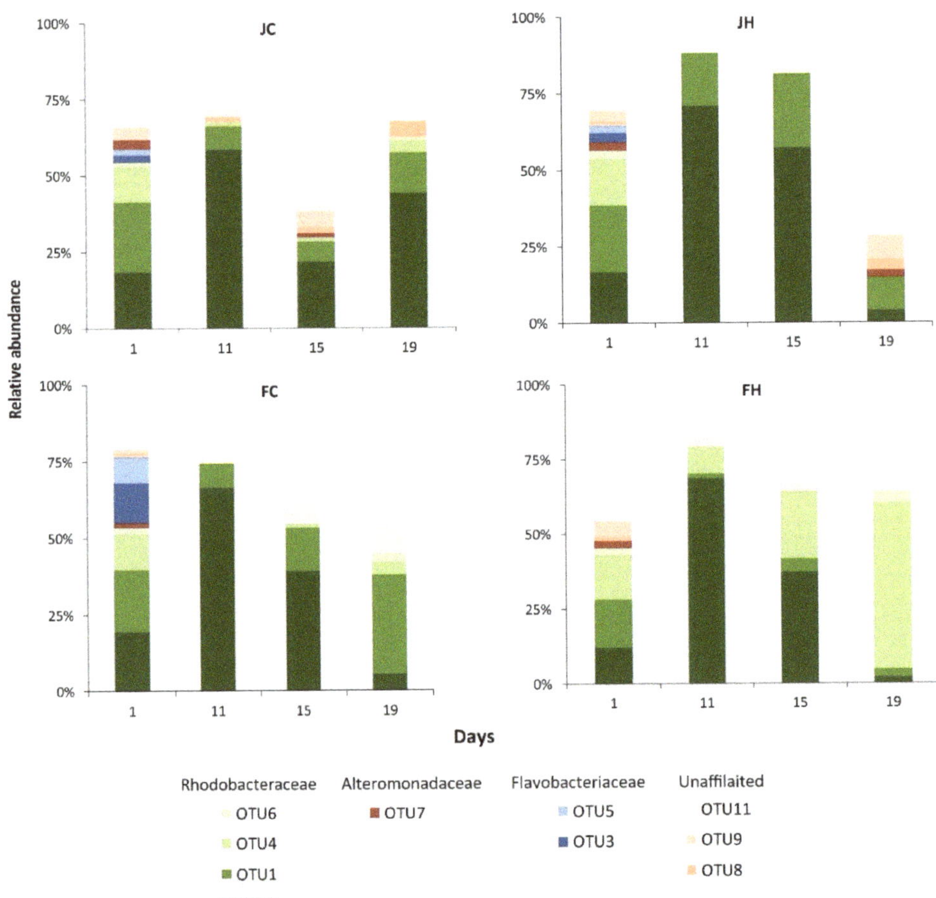

Figure 3. Temporal variation of the relative abundance of the 10 most abundant bacterial OTUs (>70% of the whole dataset). J: January, F: February, C: control (baseline light intensity), H: high light intensity.

The relative abundance of only two OTUs showed statistically significant correlations with the biomass of a specific phytoplankton species. OTU1 was positively correlated ($p = 0.007$, $r^2 = 0.648$)

with the biomass of *Chaetoceros decipiens*, while OTU2 was negatively correlated ($p = 0.017$, $r^2 = 0.587$) with the same organism.

4. Discussion

We investigated changes of heterotrophic prokaryoplankton community composition during experimentally induced coastal phytoplankton blooms, based on OTU diversity determined by 454 pyrosequencing of the bacterial V1–V3 and archaeal V3–V9 16S rRNA regions. In particular, we aimed to detect whether changes in community composition are uniform, or if a divergence in OTUs occurs across different treatments. True replication [41] was not feasible at this time due to experimental design limitations (e.g., removal of excess water could disturb the mesocosms' stability), but our analysis included four different types of phytoplankton succession induced by different light regimes. However, considering that Sommer et al. [19] did not find clear differences between replicated mesocosms for various biological parameters studied (e.g., phytoplankton biomass, relative biomass of phytoplankton groups, relative abundance of nauplii) but only between treatments, we believe that although we cannot make strong statistical conclusions, our results give insights into microbial processes that occur under different light conditions. It should be mentioned that in the January mesocosms, a phytoplankton bloom developed only in the treatment with the increased irradiance, whereas in the February mesocosms a phytoplankton bloom also developed in the control (Figure S3). Positive phytoplankton growth rates also occurred in the January control mesocosm, but their values, as well as the phytoplankton biomass, were far below the limits characterizing a bloom event [19]. For most of the analyzed samples, the sequencing depth (Figure S1) was satisfactory for revealing at least the dominant OTUs, and here we focus on these dominant operational taxonomic units (OTUs) and not at revealing the exhaustive species richness of the mesocosms.

The decrease of unique OTUs during the course and/or towards the end of the growth phase in all mesocosms depicts that some kind of selection took place at the OTU level, resulting in an increase of dominance [42]. The most frequent and abundant OTUs (i.e., those found in all mesocosms and all sampling points), belonged to Rhodobacteraceae. This taxon includes some of the most important marine Bacteria, like the *Roseobacter* clade, which is known to have a positive response to increases in chlorophyll a [43,44] and is associated with phytoplankton blooms in marine waters [7,45,46]. Additionally, *Roseobacter* spp. prevail in meso-eutrophic systems [47]. The Rhodobacteraceae-related OTU2 found in our study persisted in all mesocosms despite the different light regimes and subsequent changes in the plankton community. The peak of this OTU coincided with the initiation or peak of the phytoplankton biomass (Figures S3 and S4). It is likely that this OTU represents a bacterium that not only outcompetes the rest of the bacterial OTUs, but is also resistant to grazing pressure (not investigated here). The metabolic plasticity of several members of the Rhodobacteraceae [23,48,49] could support a bottom-up advantage for these microorganisms by using multiple and/or complex organic carbon resources and, thus, benefit from phytoplankton blooms [3,8,50] like the ones that occurred in our mesocosm. Actually, between the onset of the phytoplankton bloom and the end of the incubation (post-bloom period), DOC increasingly reached a factor of >3 [51], suggesting that bulk carbon sources were not limiting. However, further research is necessary in order to investigate the impact of different DOC compounds on specific Bacteria [52,53].

The second most frequent and abundant microorganism was OTU1, also a member of Rhodobacteraceae. However, although its initial relative abundance was similar to that of OTU2, it increased only in February mesocosms at the end of the experiment when phytoplankton biomass decreased (post-bloom period with senescent and dead phytoplankton cells) and phosphate concentration minima occurred (Figure S3). This pattern indicates a possible suppression of OTU1 abundance by OTU2 and a possible antagonistic relationship between the two microorganisms. OTU1 was a better competitor under the low phosphate concentrations recorded in February mesocosms. Furthermore, these two OTUs were the only ones that showed statistically significant correlations with the biomass of a specific phytoplankter. In particular, OTU1 was positively and OTU2 was negatively correlated to the *C. decipiens* biomass. Since Bacteria species may use DOM originating from multiple phytoplankton species [44,54] and other sources [52], these statistically significant correlations should not be considered as one-on-one relationships but might indicate the differential impact of phytoplankton exudates to specific Bacteria. Similarly to the OTU2 and OTU1 correlations, contradictory responses to phytoplankton exudates of two Rhodobacteraeae strains have also been observed previously [55]. Finally, a third type of OTU relative abundance response was common for the rest of the nine dominant OTUs. The abundance of these microorganisms always remained low, as has been the case for many bacteria in several mesocosm experiments, mostly due to competition [5].

Other important bacterial families included were two that are well known from the marine environment, Flavobacteriaceae and Alteromonadaceae, which have also been found in previous Baltic Sea mesocosm experiments [29,56]. Flavobacteriaceae have been found to respond to enrichment in particulate or dissolved organic carbon [42]. However, in the current experiment their relative abundance increased only in the January treatments. We suggest that in the February mesocosms, Actinomycetales acted antagonistically to Flavobacteriaceae and played a significant role in the breakdown and recycling of organic matter [57]. The lag of the Alteromonadaceae-related OTU relative abundance in all treatments might have been related to their grazing susceptibility, as has been suggested in previous mesocosm experiments [8] and natural marine environments [58]. Interestingly, a previous study [29] using CARD-FISH found a Gammaproteobacteria-related OTU (*Glaciecola*) to dominate during diatom blooms in low temperature (< 8 °C) mesocosm experiments with whole plankton communities in the Baltic Sea. According to these authors, dominant diatom (*S. costatum* and *Thalassiosira rotula)* exudates favored *Glaciecola* spp. In our experiment, we found a *Glaciecola*-related OTU (OTU7) in all mesocosms, but it never proliferated (even though *S. costatum* was one of the dominant phytoplankters) [19]. *S. costatum sensu lato* harbors cryptic species [59], and this feature could probably explain the discrepancies between studies. Members of Verrucomicrobia also occurred in noticeable relative abundance towards the end of the phytoplankton blooms in the FH mesocosms. Verrucomicrobia are found in freshwater blooms of eutrophic systems [60,61], while a recent report couples the constant presence of Verrucomicrobia with enzymatic degradation of polysaccharides in Arctic waters [62], inferring a tight link of this group to organic matter originating from phytoplankton blooms. Although direct comparisons of the BCS between our results and those of other mesocosm studies from the Baltic Sea [29,56] and other aquatic habitats [5,8,63] are not directly feasible due to different experimental setups and fingerprinting methods used (e.g., tag pyrosequencing vs DGGE), some common major groups, such as those at the family level, can been seen. Overall, the dynamics of different families hint at the competitive advantage of Rhodobacteraceae against Flavobacteriaceae and Alteromonadaceae in our mesocosms, which is probably related to their resistance to grazing as well as to their ability to use multiple and/or complex organic carbon resources, as stated earlier.

In order to investigate if and how the associations of the 11 OTUs found in all mesocosms changed as a result of the different treatments, we used association network analysis of co-occurrence patterns. We found that the interrelationships of these OTUs varied between the four different mesocosms (OTU6 did not show any correlations, see Figures S5 and S6). For example, the high-light period in February mesocosms led to higher growth and production rates for phytoplankton, as well as to a higher peak of biomass [19]. Under these conditions, only a few nodes (OTU7, OTU9, OTU20) made most of the interactions, presumably as a result of being the ones best adapted to the most specific conditions of the highest phytoplankton biomass and irradiance, and because several other OTUs need to associate with them to grow. During DOC degradation, intermediate compounds are formed [52,64], providing more available niches [48] that could be occupied by a variety of OTUs (Figure 3). It seems, thus, that as light increases, more specific bacterial relationships are formed in terms of OTU relative abundance.

Although it is known that whole prokaryoplankton communities in the Baltic Sea are responsive to climate-induced changes [65,66], it remains unknown whether such responses are due to specific taxa of bacteria and/or archaea. Moreover, the effect of temperature increase on bacteria community composition can be direct or indirect [29], but far less is known of the effect of changes in light intensity and duration [51]. In this study, we found that specific bacteria had different responses in terms of OTU abundance in mesocosms under in situ and high light intensities, and also that an early phytoplankton bloom [19] can result in a different bacterial community structure. Since different light conditions resulted in different phytoplankton biomass and community structure patterns, further investigation is needed based on factorial experimental design and replication to make safe conclusions about the "significance" of different parameters on BCS. Our study shows that just as the distinct members of phytoplankton communities have responded distinctively in successive studies induced by simulated climate change effects, prokaryoplankton should not be considered as a "black box", since it also exhibits distinct temporal patterns and interrelationships as a result of such environmental changes.

Supplementary Materials: The following are available online at http://www.mdpi.com/1424-2818/11/10/195/s1: Figure S1. Rarefaction curves of Operational Taxonomic Units (OTUs) numbers against the number of high-quality reads, at ≥97 similarity, in the studied mesocosms; Figure S2. Maximum likelihood phylogenetic tree of the Bacteria most dominant operational taxonomic units (OTU) and the ones found in all four mesocosms all sampling points. Figure S3. Temporal dynamics of phosphate, nitrate to total dissolved inorganic nitrogen (DIN) ration and phytoplankton biomass in the studied mesocosms; Figure S4. Taxonomic composition of the most abundant (>1%) bacterial OTUs in the studied mesocosms. Figure S5. Association network diagram of Spearman's correlations (edges) between the 10 operational taxonomic units (numbers in nodes) shared among the four mesocosms (yellow nodes); Table S1. Relative abundance (%) of the most abundant (≥1%) OTUs. J: January, F: February, C: control (baseline light intensity), H: high light intensity.

Author Contributions: Conceptualization, K.A.K. and U.S.; Data curation, H.K. and K.A.K.; Formal analysis, H.K., K.A.K. and U.S.; Funding acquisition, K.A.K. and U.S.; Methodology, H.K., K.A.K, M.M.-G. and U.S.; Writing—original draft, H.K. and K.A.K.; Writing—review & editing, H.K., K.A.K., M.M.-G. and U.S.

Acknowledgments: "This research was funded by European Union Seventh Framework Program (FP7/2007–2013), grant number 228224, MESOAQUA" Antje Biermann is fully acknowledged for sharing DOC measurements, and Antonis Michas is acknowledged for assistance during DNA extraction.

Conflicts of Interest: The authors declare no conflict of interest.

References

1. Tada, Y.; Nakaya, R.; Goto, S.; Yamashita, Y.; Suzuki, K. Distinct bacterial community and diversity shifts after phytoplankton-derived dissolved organic matter addition in a coastal environment. *J. Exp. Mar. Biol. Ecol.* **2017**, *495*, 119–128. [CrossRef]
2. Fenchel, T. Biogeography for Bacteria. *Science* **2003**, *301*, 925–926. [CrossRef] [PubMed]
3. Hanson, C.A.; Fuhrman, J.A.; Horner-Devine, M.C.; Martiny, J.B.H. Beyond biogeographic patterns: Processes shaping the microbial landscape. *Nat. Rev. Microbiol.* **2012**, *10*, 497–506. [CrossRef] [PubMed]
4. Thorton, D.C.O. Dissolved organic matter (DOM) release by phytoplankton in the contemporary and future ocean. *Eur. J. Phycol.* **2014**, *49*, 20–46. [CrossRef]
5. Paver, S.F.; Hayek, K.R.; Crabb, D.B.; Davis-Richardson, A.G.; Rosario-Passapera, R.; Gano, K.A.; Fagen, J.R.; Brown, C.T.; Giongo, A.; Triplett, E.W.; et al. Interactions between specific phytoplankton and bacteria affect lake bacterial community succession. *Environ. Microbiol.* **2013**, *15*, 2489–2504. [CrossRef]
6. Bagatini, I.L.; Eiler, A.; Bertilsson, S.; Klaveness, D.; Tessarolli, L.P.; Vieira, A.A.H. Host-Specificity and Dynamics in Bacterial Communities Associated with Bloom-Forming Freshwater Phytoplankton. *PLoS ONE* **2014**, *9*, e85950. [CrossRef]
7. Pinhassi, J.; Winding, A.; Binnerup, S.; Zweifel, U.; Riemann, B.; Hagström, A. Spatial variability in bacterioplankton community composition at the Skagerrak-Kattegat Front. *Mar. Ecol. Prog. Ser.* **2003**, *255*, 1–13. [CrossRef]
8. Tada, Y.; Taniguchi, A.; Sato-Takabe, Y.; Hamasaki, K. Growth and succession patterns of major phylogenetic groups of marine bacteria during a mesocosm diatom bloom. *J. Oceanogr.* **2012**, *68*, 509–519. [CrossRef]
9. Zhou, J.; Richlen, M.L.; Sehein, T.R.; Kulis, D.M.; Anderson, D.M.; Cai, Z. Microbial Community Structure and Associations During a Marine Dinoflagellate Bloom. *Front. Microbiol.* **2018**, *9*, 1201. [CrossRef]
10. Meziti, A.; Kormas, K.A.; Moustaka-Gouni, M.; Karayanni, H. Spatially uniform but temporally variable bacterioplankton in a semi-enclosed coastal area. *Syst. Appl. Microbiol.* **2015**, *38*, 358–367. [CrossRef]

11. Hoppe, H.G.; Giesenhagen, H.C.; Köppe, R.; Hansen, H.P.; Göcke, K. Impact of change in climate and policy from 1988 to 2007 on environmental and microbial variables at the time series station Boknis Eck, Baltic Sea. *Biogeosciences* **2013**, *10*, 4529–4546. [CrossRef]
12. Tsiola, A.; Tsagaraki, T.M.; Giannakourou, A.; Nikolioudakis, N.; Yücel, N.; Herut, B.; Pitta, P. Bacterial Growth and Mortality after Deposition of Saharan Dustand Mixed Aerosols in the Eastern Mediterranean Sea: A Mesocosm Experiment. *Front. Mar. Sci.* **2017**, *3*, 281. [CrossRef]
13. Stefanidou, N.; Genitsaris, S.; Lopez-Bautista, J.; Sommer, U.; Moustaka-Gouni, M. Effects of heat shock and salinity changes on coastal Mediterranean phytoplankton in a mesocosm experiment. *Mar. Biol.* **2018**, *165*, 154. [CrossRef]
14. Shade, A.; Read, J.S.; Youngblut, N.D.; Fierer, N.; Knight, R.; Kratz, T.K.; Lottig, N.R.; Roden, E.E.; Stanley, E.H.; Stombaugh, J.; et al. Lake microbial communities are resilient after a whole-ecosystem disturbance. *ISME J.* **2012**, *6*, 2153–2167. [CrossRef] [PubMed]
15. Allison, S.D.; Martiny, J.B.H. Resistance, resilience, and redundancy in microbial communities. *Proc. Natl. Acad. Sci. USA* **2008**, *105*, 11512–11519. [CrossRef] [PubMed]
16. Downing, A.L.; Leibold, M.A. Species richness facilitates ecosystem resilience in aquatic food webs. *Freshw. Biol.* **2010**, *55*, 2123–2137. [CrossRef]
17. Ruiz-González, C.; Simó, R.; Sommaruga, R.; Gasol, J.M. Away from darkness: A review on the effects of solar radiation on heterotrophic bacterioplankton activity. *Front. Microbiol.* **2013**, *4*, 131. [CrossRef]
18. Yamamichi, M.; Kazama, T.; Tokita, K.; Katano, I.; Doi, H.; Yoshida, T.; Hairston, N.G., Jr.; Urabe, J. A shady phytoplankton paradox: When phytoplankton increases under low light. *Proc. R. Soc. B* **2018**, *285*, 20181067. [CrossRef]
19. Sommer, U.; Lengfellner, K.; Lewandowska, A. Experimental induction of a coastal spring bloom early in the year by intermittent high-light episodes. *Mar. Ecol. Prog. Ser.* **2012**, *446*, 61–71. [CrossRef]
20. Paver, S.F.; Kent, A.D. Direct and Context-dependent effects of light, temperature, and phytoplankton shape bacterial community composition. *Ecosphere* **2017**, *8*, e01948. [CrossRef]
21. Alonso-Sáez, L.; Gasol, J.M.; Lefort, T.; Hofer, J.; Sommaruga, R. Effect of Natural Sunlight on Bacterial Activity and Differential Sensitivity of Natural Bacterioplankton Groups in Northwestern Mediterranean Coastal Waters. *Appl. Environ. Microbiol.* **2006**, *72*, 5806–5813. [CrossRef] [PubMed]
22. Ruiz-González, C.; Lefort, T.; Massana, R.; Simó, R.; Gasol, J.M. Diel changes in bulk and single-cell bacterial heterotrophic activity in winter surface waters of the northwestern Mediterranean Sea. *Limnol. Oceanogr.* **2012**, *57*, 29–42. [CrossRef]
23. Sánchez, O.; Koblížek, M.; Gasol, J.M.; Ferrera, I. Effects of grazing, phosphorus and light on the growth rates of major bacterioplankton taxa in the coastal NW Mediterranean. *Environ. Microbiol. Rep.* **2017**, *9*, 300–309. [CrossRef] [PubMed]
24. Schwalbach, M.S.; Brown, M.; Fuhrman, J.A. Impact of light on marine bacterioplankton community structure. *Aquat. Microb. Ecol.* **2005**, *39*, 235–245. [CrossRef]
25. Agogué, H.; Joux, F.; Obernosterer, I.; LeBaron, P. Resistance of Marine Bacterioneuston to Solar Radiation. *Appl. Environ. Microbiol.* **2005**, *71*, 5282–5289. [CrossRef]
26. Schlesinger, W.H. *Biogeochemistry: Treatise on Geochemistry*, 1st ed.; Elsevier: Amsterdam, The Netherlands, 2005; Volume 8.
27. Parker, M.S.; Armbrust, E.V. Synergistic effects of light, temperature, and nitrogen source on transcription of genes for carbon and nitrogen metabolism in the centric diatom *Thalassiosira pseudonana* (Bacillariophyceae). *J. Phycol.* **2005**, *41*, 1142–1153. [CrossRef]
28. Blanchet, M.; Fernandez, C.; Joux, F. Photoreactivity of riverine and phytoplanktonic dissolved organic matter and its effects on the dynamics of a bacterial community from the coastal Mediterranean Sea. *Prog. Oceanogr.* **2018**, *163*, 82–93. [CrossRef]
29. Von Scheibner, M.; Dörge, P.; Biermann, A.; Sommer, U.; Hoppe, H.G.; Jürgens, K. Impact of warming on phyto-bacterioplankton coupling and bacterial community composition in experimental mesocosms. *Environ. Microbiol.* **2014**, *16*, 718–733. [CrossRef]
30. Dowd, S.; Callaway, T.R.; Wolcott, R.D.; Sun, Y.; McKeehan, T.; Hagevoort, R.G.; Edrington, T.S. Evaluation of the bacterial diversity in the feces of cattle using 16S rDNA bacterial tag-encoded FLX amplicon pyrosequencing (bTEFAP). *BMC Microbiol.* **2008**, *8*, 125. [CrossRef]

31. Schloss, P.D.; Westcott, S.L.; Ryabin, T.; Hall, J.R.; Hartmann, M.; Hollister, E.B.; Lesniewski, R.A.; Oakley, B.B.; Parks, D.H.; Robinson, C.J.; et al. Introducing mothur: Open-Source, Platform-Independent, Community-Supported Software for Describing and Comparing Microbial Communities. *Appl. Environ. Microbiol.* **2009**, *75*, 7537–7541. [CrossRef]
32. Quince, C.; Lanzén, A.; Curtis, T.P.; Davenport, R.J.; Hall, N.; Head, I.M.; Read, L.F.; Sloan, W.T. Accurate determination of microbial diversity from 454 pyrosequencing data. *Nat. Methods* **2009**, *6*, 639–641. [CrossRef]
33. Edgar, R.C. Search and clustering orders of magnitude faster than BLAST. *Bioinformatics* **2010**, *26*, 2460–2461. [CrossRef] [PubMed]
34. Reeder, J.; Knight, R. The 'rare biosphere': A reality check. *Nat. Rev. Microbiol.* **2009**, *6*, 636–637. [CrossRef] [PubMed]
35. Behnke, A.; Engel, M.; Christen, R.; Nebel, M.; Klein, R.R.; Stoeck, T. Depicting more accurate pictures of protistan community complexity using pyrosequencing of hypervariable SSU rRNA gene regions. *Environ. Microbiol.* **2010**, *13*, 340–349. [CrossRef] [PubMed]
36. Kunin, V.; Engelbrektson, A.; Ochman, H.; Hugenholtz, P. Wrinkles in the rare biosphere: Pyrosequencing errors can lead to artificial inflation of diversity estimates. *Environ. Microbiol.* **2010**, *12*, 118–123. [CrossRef] [PubMed]
37. Lemos, L.N.; Fulthorpe, R.R.; Triplett, E.W.; Roesch, L.F. Rethinking microbial diversity analysis in the high throughput sequencing era. *J. Microbiol. Methods* **2011**, *86*, 42–51. [CrossRef] [PubMed]
38. Pruesse, E.; Quast, C.; Knittel, K.; Fuchs, B.M.; Ludwig, W.; Peplies, J.; Glöckner, F.O. SILVA: A comprehensive online resource for quality checked and aligned ribosomal RNA sequence data compatible with ARB. *Nucleic Acids Res.* **2007**, *35*, 7188–7196. [CrossRef]
39. Hammer, Ø.; Harper, D.A.T.; Ryan, P.D. PAST: Paleontological statistics software package for education and data analysis. *Palaeontol. Electron.* **2001**, *4*, 9.
40. Smoot, M.E.; Ono, K.; Ruscheinski, J.; Wang, P.L.; Ideker, T. Cytoscape 2.8: New features for data integration and network visualization. *Bioinformatics* **2011**, *27*, 431–432. [CrossRef]
41. Prosser, J.I. Replicate or lie. *Environ. Microbiol.* **2010**, *12*, 1806–1810. [CrossRef]
42. Karayanni, H.; Meziti, A.; Spatharis, S.; Genitsaris, S.; Courties, C.; Kormas, K.A. Changes in Microbial (Bacteria and Archaea) Plankton Community Structure after Artificial Dispersal in Grazer-Free Microcosms. *Microorganisms* **2017**, *5*, 31. [CrossRef] [PubMed]
43. Brinkhoff, T.; Giebel, H.A.; Simon, M. Diversity, ecology, and genomics of the Roseobacter clade: A short overview. *Arch. Microbiol.* **2008**, *189*, 531–539. [CrossRef] [PubMed]
44. Lefort, T.; Gasol, J.M. Global-scale distributions of marine surface bacterioplankton groups along gradients of salinity, temperature, and chlorophyll: A meta-analysis of fluorescence in situ hybridization studies. *Aquat. Microb. Ecol.* **2013**, *70*, 111–130. [CrossRef]
45. Fandino, L.; Riemann, L.; Steward, G.; Long, R.; Azam, F. Variations in bacterial community structure during a dinoflagellate bloom analyzed by DGGE and 16S rDNA sequencing. *Aquat. Microb. Ecol.* **2001**, *23*, 119–130. [CrossRef]
46. Zubkov, M.V.; Fuchs, B.M.; Archer, S.D.; Kiene, R.P.; Amann, R.; Burkill, P.H. Linking the composition of bacterioplankton to rapid turnover of dissolved dimethylsulphoniopropionate in an algal bloom in the North Sea. *Environ. Microbiol.* **2001**, *3*, 304–311. [CrossRef] [PubMed]
47. Fuhrman, J.A.; Hagström, A. Bacterial and archaeal community structure and its patterns. In *Microbial Ecology of the Oceans*, 2nd ed.; Kirchman, D.L., Ed.; Wiley-Blackwell: Hoboken, NJ, USA, 2008; pp. 45–90.
48. Newton, R.J.; Griffin, L.E.; Bowles, K.M.; Meile, C.; Gifford, S.; Givens, C.E.; Howard, E.C.; King, E.; Oakley, C.A.; Reisch, C.R.; et al. Genome characteristics of a generalist marine bacterial lineage. *ISME J.* **2010**, *4*, 784–798. [CrossRef] [PubMed]
49. Brenner, D.J.; Krieg, N.R.; Staley, J.T. The Proteobacteria: Part C: The Alpha-, Beta-, Delta-, and Epsilonproteobacteria. In *Bergey's Manual of Systematic Bacteriology*, 2nd ed.; Garriy, G.M., Ed.; Springer: New York, NY, USA, 2005; pp. 161–229.
50. Allers, E.; Pinhassi, J.; Gasol, J.M.; Šimek, K.; Pernthaler, J.; Gómez-Consarnau, L.; Gómez-Consarnau, L. Response of Alteromonadaceae and Rhodobacteriaceae to glucose and phosphorus manipulation in marine mesocosms. *Environ. Microbiol.* **2007**, *9*, 2417–2429. [CrossRef]
51. Biermann, A.; Engel, A.; Riebesell, U. Changes in organic matter cycling in a plankton community exposed to warming under different light intensities. *J. Plankton Res.* **2014**, *36*, 658–671. [CrossRef]

52. Hansell, D.A.; Carlson, C.A. *Biogeochemistry of Marine Dissolved Organic Matter*, 1st ed.; Academic Press: San Diego, CA, USA, 2002.
53. Jiao, N.; Herndl, G.J.; Hansell, D.A.; Benner, R.; Kattner, G.; Wilhelm, S.W.; Kirchman, D.L.; Weinbauer, M.G.; Luo, T.; Chen, F.; et al. Microbial production of recalcitrant dissolved organic matter: Long-term carbon storage in the global ocean. *Nat. Rev. Microbiol.* **2010**, *8*, 593–599. [CrossRef]
54. Sarmento, H.; Gasol, J.M. Use of phytoplankton-derived dissolved organic carbon by different types of bacterioplankton. *Environ. Microbiol.* **2012**, *14*, 2348–2360. [CrossRef]
55. Grossart, H.; Simon, M. Interactions of planktonic algae and bacteria: Effects on algal growth and organic matter dynamics. *Aquat. Microb. Ecol.* **2007**, *47*, 163–176. [CrossRef]
56. Lindh, M.V.; Riemann, L.; Baltar, F.; Romero-Oliva, C.; Salomon, P.S.; Granéli, E.; Pinhassi, J. Consequences of increased temperature and acidification on bacterioplankton community composition during a mesocosm spring bloom in the Baltic Sea. *Environ. Microbiol. Rep.* **2013**, *5*, 252–262. [CrossRef]
57. Subramani, R.; Aalbersberg, W. Marine actinomycetes: An ongoing source of novel bioactive metabolites. *Microbiol. Res.* **2012**, *167*, 571–580. [CrossRef] [PubMed]
58. Beardsley, C.; Pernthaler, J.; Wosniok, W.; Amann, R. Are Readily Culturable Bacteria in Coastal North Sea Waters Suppressed by Selective Grazing Mortality? *Appl. Environ. Microbiol.* **2003**, *69*, 2624–2630. [CrossRef] [PubMed]
59. Smayda, T.J. Cryptic planktonic diatom challenges phytoplankton ecologists. *Proc. Natl. Acad. Sci. USA* **2011**, *108*, 4269–4270. [CrossRef]
60. Ye, W.; Liu, X.; Lin, S.; Tan, J.; Pan, J.; Li, D.; Yang, H. The vertical distribution of bacterial and archaeal communities in the water and sediment of Lake Taihu. *FEMS Microbiol. Ecol.* **2009**, *70*, 263–276. [CrossRef] [PubMed]
61. Kormas, K.A.; Vardaka, E.; Moustaka-Gouni, M.; Kontoyanni, V.; Petridou, E.; Gkelis, S.; Neofitou, C. Molecular detection of potentially toxic cyanobacteria and their associated bacteria in lake water column and sediment. *World J. Microbiol. Biotechnol.* **2010**, *26*, 1473–1482. [CrossRef]
62. Cardman, Z.; Arnosti, C.; Durbin, A.; Ziervogel, K.; Cox, C.; Steen, A.D.; Teske, A. Verrucomicrobia Are Candidates for Polysaccharide-Degrading Bacterioplankton in an Arctic Fjord of Svalbard. *Appl. Environ. Microbiol.* **2014**, *80*, 3749–3756. [CrossRef]
63. Teira, E.; Gasol, J.M.; Aranguren-Gassis, M.; Fernández, A.; González, J.; Lekunberri, I.; Aranguren-Gassis, M.; Fernández-Carrera, A.; Álvarez-Salgado, X.A.; Aranguren-Gassis, M.; et al. Linkages between bacterioplankton community composition, heterotrophic carbon cycling and environmental conditions in a highly dynamic coastal ecosystem. *Environ. Microbiol.* **2008**, *10*, 906–917. [CrossRef]
64. Murray, A.; Arnosti, C.; De La Rocha, C.; Grossart, H.; Passow, U. Microbial dynamics in autotrophic and heterotrophic seawater mesocosms. II. Bacterioplankton community structure and hydrolytic enzyme activities. *Aquat. Microb. Ecol.* **2007**, *49*, 123–141. [CrossRef]
65. Nydahl, A.; Panigrahi, S.; Wikner, J. Increased microbial activity in a warmer and wetter climate enhances the risk of coastal hypoxia. *FEMS Microbiol. Ecol.* **2013**, *85*, 338–347. [CrossRef] [PubMed]
66. Panigrahi, S.; Nydahl, A.; Anton, P.; Wikner, J. Strong seasonal effect of moderate experimental warming on plankton respiration in a temperate estuarine plankton community. *Estuar. Coast. Shelf Sci.* **2013**, *135*, 269–279. [CrossRef]

© 2019 by the authors. Licensee MDPI, Basel, Switzerland. This article is an open access article distributed under the terms and conditions of the Creative Commons Attribution (CC BY) license (http://creativecommons.org/licenses/by/4.0/).